Panzerjaeger Evolution

The concept of creating a mobile counter-striking force, armed with anti-tank guns mounted on full-tracked chassis, was introduced long before the Nazi era of aggression. Already in the Spring of 1927, General von Blomberg of the Truppenamt stated that motorized towing vehicles and Raupenselbstfahrlafette (full-tracked self-propelled carriages) were the real goal of the anti-tank program. By July 1927, they had already come to the realization that: "*Tankzerstoerer (tank destroyers) are the most effective weapon against tanks. These should be full-tracked or wheel-cum-track self-propelled guns with the lowest possible height and highest possible speed. They must take advantage of the tactical weakness of a tank attack by mobile fighting - if possible striking in the flank or rear. This will always force a tank attack to turn aside.*"

Due to the dire economic conditions of the period, money wasn't available for an extensive research and development program. A quick and cheap interim solution was needed. Guns were simply mounted on commercially available full-tracked chassis (like bulldozers). After purchasing and testing a few trial vehicles, this effort was abandoned. It was concluded that this Zwischenloesung (interim solution) didn't meet the required military specifications for cross-country mobility and low-profile. They recognized that a Tankzerstoerer needed to have a full-tracked chassis that was specifically designed to meet specifications derived from tactical requirements.

In the Fall of 1927, Krupp embarked on the development of just such a full-tracked chassis designed for a self-propelled gun. Initially known as a Motorlafette (motor carriage), the name was changed later to leichte Selbstfahrkanone (light self-propelled gun), abbreviated as L.S.K. The first chassis had a very elaborate suspension with rubber band tracks - even the sprocket and idler pivoted when crossing obstacles. It failed and was quickly sent back to Krupp for major modification including a simpler suspension. An attempt was also made to adapt the Leichttraktor (code name for their light tank) as a self-propelled gun by mounting the 3.7 cm Tak L/45 in a fully traversable turret. Both projects were tested and abandoned.

Next, faith was placed in the correct solution being half-tracked chassis, which had been successfully developed by Herr Kniepkamp as artillery prime movers. By the Fall of 1935, specifications had been created to govern the design of these new half-tracked Panzerjaeger. The key requirements were higher speed, better cross-country mobility, and lower cost per unit than tanks. Two versions were developed - a lighter Selbstfahrlafette armed with a 3.7 cm Pak and a heavier model armed with a 7.5 cm Kanone L/40 capable of penetrating the armor of any known enemy tank. By the start of the war, developmental work had ceased on the lighter 3.7 cm Sfl. but work continued into 1941 on the heavier 7.5 cm Sfl. L/40.8. Only two of the final model Pz.Sfl.II were built, tested in North Africa, and then the program was abandoned. These half-tracked vehicles had failed to achieve any of the key requirements. They weren't faster, didn't have better cross-country mobility, nor were they cheaper to produce than tanks. Until the E-10 "Hetzer" project (refer to Panzer Tracts 20-1), the Pz.Sfl.II was the last attempt by the Heereswaffenamt to design a chassis specifically for a cheap, high-speed Panzerjaeger.

With the exception of a few Schartenknacker (bunker busters), developed to attack Maginot line fortifications, being used in an anti-tank role, during the war Panzer-Jaeger-Abteilungen (tank destroyer battalions) had to be content with make-shift self-propelled guns as a replacement for towed anti-tank guns. Most of these make-shift Panzerjaeger were simply standard anti-tank guns bolted onto the top of available tank chassis and protected on the front and sides by thin armor shields.

In addition to several 8.8 cm Waffentraeger projects at the end, there were 23 of these make-shift Panzerjaeger-far too many to be adequately covered in one Panzer-Tracts volume. This first volume (Panzer Tracts 7-1) covers Tankzerstoerer/Panzerjaeger development, production, and frontline employment from 1927 to 1941. Future volumes in this series will cover the Panzerjaeger conceived and produced from 1942 to 1945.

The highest level of historical knowledge is achieved when one gains insight into the decisions behind events—an understanding of why decisions were made—not just what happened. Thanks to the British School of Tank Technology, the development records for some Krupp projects survived. Compare the detailed accounts for the Krupp L.S.K. and Pz.Sfl.IVa projects with the total lack of any information on what occurred during the evolution of Rheinmetall projects.

Fabulous insight into the strengths and weaknesses of Panzerjaeger in action in 1940/41 is gained from the few surviving experience reports written at the time by Panzer-Jaeger unit commanders. One commander didn't appreciate the better viewing capability from an open-top shield. Statements made by one commander on the results of firing the 10.5 cm Kanone Sfl. at long range reveal that the reality of battlefield capabilities (1000 meters as the maximum range to engage targets) were far different than simplistic conclusions derived from data evaluation.

Tak-Zwischenloesung for the first Tankzerstoerer

3.7 cm Tak Specifications

On 13 February 1926, Inspektion fuer Waffen und Geraet Apt.4 created the design specifications for a Sonderlafette fuer das 3,7 cm Rohr L/45 des l.M.W. (Tak) (special carriage for a towed 3.7 cm anti-tank gun). These specifications were sent to Krupp with a request for a design proposal with the warning that this information was to be protected with the strictest secrecy.

Specifications for the design of a Sonderlafette fuer das 3,7 cm Rohr L/45 des l.M.W. (Tak)

The primary purpose of the gun is as a Tankabwehrgeschuetz (anti-tank gun). It should penetrate the best 40-mm-thick armor plates (set at an angle of 30 degrees from vertical) at a range of 500 meters with tracer shells and have sufficient fragmentation to knock the crews out of action behind the armor. In exceptional cases it should also be used against ground targets such as machineguns protected by steel shields. Engaging low-flying aircraft is not a consideration. It must be able to follow infantry in all types of terrain and be easily maneuverable. The crew of one gun commander and two men (initially one man) must quickly bring the gun into action from cover.

The 3.7 cm Rohr L/45 of the l.M.W. is available. The gun tube should remain interchangeable with the l.M.W. mount. An armor-piercing projectile with tracer weighing 0.6 kg is fired with a muzzle velocity of 760 m/sec. A muzzle brake is desired. The breech is a semi-automatic standard breech block with right and left operation. It has a simple recoil cylinder with a "Paier" recuperator.

The wheeled carriage for horse or crew towing should be capable of disassembly for transport by men or pack animals. The long trails prevent the gun from skidding when fired at 0 degrees. Elevation arc is -5 to +15 degrees with traverse of at least 20 degrees to each side. The firing height is to be kept as low as possible while mounted on the wheeled carriage. The ground clearance while traveling should not be under 250 mm and it should have a wheelbase of 1125 mm width.

The gun shield (about 4.5 mm thick) should stop SS (7.92 mm lead bullets) rounds from penetrating at 500 meters range. Total weight in firing position should be less than 250 kg. The aiming device is the Rundblickfernrohr developed for the Gebirgs-Geschuetz in addition to the normal Kimme und Korn (front and rear open sight). Loops should be located on the outside for fastening camouflage. One or two ammunition boxes should hold at least 10 rounds.

Raupenselbstfahrlafette

Decisions concerning the choice of Tankabwehrwaffe (anti-tank weapons) for the Reichswehr

(German army) were recorded in a report signed by General von Blomberg at a Truppenamt meeting held on 19 May 1927. The conclusions resulting from these discussions were:

Stick with the current Inspektorat 2 plans for a 3.7 cm Tankabwehrkanone (abbreviated Tak) based on:
a. the far advanced development of the gun and ammunition will be the quickest way to reach the goal of procuring a Tankabwehrwaffe for the infantry for employment in the front battle zone,
b. the penetration capability of the 3.7 cm Tak is completely adequate for defeating the armor of known light and medium tanks,
c. the mobility of the weapon, the desired low firing height, and potential for sufficient on-hand ammunition is improved by the small (3.7 cm) caliber, and
d. lowest procurement cost.

The decision to accept the 3.7 cm Tak as the only Tankabwehrwaffe remains supportable even considering the known disadvantages:
a. little suitability for other combat employment,
b. differences of opinion on the final carriage design,
c. selecting a larger caliber to counter a major army outfitted with modern tanks.

It was decided not to develop other larger caliber anti-tank guns. Intermediate calibers were rejected for combating heavy tanks. Strive to develop a standard 7.5 cm caliber self-propelled gun.

Motorized Schlepper (towing vehicles) or Raupenselbstfahrlafette (full-tracked self-propelled carriages) are the goal for the 3.7 cm Tak. There are insufficient data at present to determine which type is preferable. The necessary trials are to be started. Until the motorization question is clarified, horsedrawn carriages are planned as a stopgap solution.

Plans have been made to acquire about 300 3.7 cm Tak by 1932 within the Minenwerferprogramm. The successful accomplishment of this procurement plan will be based on the financial situation.

Employment of the 5 cm Marinebkommrohre as a Tankabwehrwaffe or Infanteriegeschuetz was rejected based on its higher costs. In addition, the necessary expenditure for conversion, mounting, and procuring ammunition wouldn't be accomplished in time.

Tankzerstoerer

On 4 July 1927, the Chef des Heereswaffenamts recorded his position on the type of Tankabwehrwaffe that was needed:

Tankzerstoerer (tank destroyers) are the most effective weapon against tanks. These should be full-tracked or wheel-cum-track self-propelled guns with the lowest possible height and highest possible speed. They must take advantage of the tactical weakness of a tank attack by mobile fighting - if possible striking in the flank or rear. This will always force the tank attack to turn aside. It appears to me that a Zerstoererbatterie of 12 Geschuetzen is needed for each division.

The 3.7 cm Kanone can be the gun mounted in the Zerstoerer. It has high penetrating power, a high rate of fire, and a larger number of rounds can be carried. Observation of shots must be possible out to a

Left:
A 7.7 cm Kanone mounted on the commercially available Hanomag WD-Schlepper 50 PS (WJS)

Right:
A 3.7 cm Tak mounted on the commercially available Hanomag WD-Schlepper 25 PS (WJS)

range of about 2000 meters. A functional tracer has still not been developed. A 3.7 cm Panzergranate without a tracer can't be observed for target corrections. If a tracer for the 3.7 cm can't be developed, the caliber must be increased until the Panzergranate can be observed. Further trials are necessary.

In my opinion, a horse-drawn 3.7 cm Tak will not be a usable weapon for the infantry. In time of crises either the guns won't be where they are needed or they will be available in such small numbers that they won't be effective against a modern tank attack.

It is proposed that a Tankzerstoerer auf Raupen be developed with a 3.7 cm gun.

Tak-Zwischenloesung

On 8 October 1926, the Inspektion fuer Waffen und Geraet Abt.4 asked Fried. Krupp A.G., Essen to provide a proposal for the design of a 7.5 cm Infanterie-Geschuetz auf Selbstfahrender Motorlafette in accordance with the following specifications:

The general requirements are for a Motorlafette mounting a 7.5 cm Infanterie-Geschuetz with dismountable carriage that is transportable by men, 20 Panzergeschosse for Tankabwehr, and two-man crew. The chassis is to be a <u>commercially available</u> Raupenschlepper (fully-tracked caterpillar tractor) or, if possible, many components taken from one. The same Schlepper with a different superstructure is to be used as a Munitions-Transportfahrzeug that will also tow an ammunition trailer. It must be possible to transfer the gun to the Munitions-Transportfahrzeug while on the battlefield. During road marches, two to four Motor-Geschuetzen or Mun.-Wagen should be loaded onto a commercially available self-propelled transport vehicle.

The gun tube, carriage, and sight are from the new 7.5 cm Gebirgs-Kanone P. Capabilities, ammunition, and loading are the same as the Gebirgs-Kanone with the exception that the anti-tank ammunition (Panzergeschosse) is to be Patronenmunition (projectile fixed in cartridges). The weapon is to be fired on the vehicle to the front. Traverse arc of the mounted gun is to be 15 degrees right and 15 degrees left of center with an elevation arc of at least -10 to +40 degrees. Gunsights are to be mounted for both direct and indirect sighting. An 8-mm-thick slanted gun shield is to protect the front and sides and have loops for camouflage.

The cross-country speed and climbing ability will be based on the type of commercially available chassis that is selected. The goal is 12 km/hr cross-country. The radiator is to be protected by 10-mm-thick armor. A hitch is needed for a dual-axle trailer.

A Munitions-Fahrzeug is to have the same chassis as the Motor-Geschuetz. Ammunition is to be stowed in baskets. The side walls of the stowage bin are to be 5-mm-thick armor. It is desirable to have a device that will dump the ammunition. A hitch is needed for a trailer. The four-wheeled trailer for ammunition transport is to be about the same length as the motorized chassis. It should have a low profile and be made out of commercially available parts.

The motorized Transportwagen for two to four tracked chassis or trailers should have a very low bed for easier loading and unloading. It is only to be a commercially available road vehicle (or made from commercially available parts) with a road speed of about 40 km/hr.

In.4 asked Krupp to develop a rough conceptual proposal so that the first meeting to discuss the design could be held at the end of November 1926. Further detailed work is to be based on the results of the discussions.

In response to the October 1926 specification, Krupp first proposed a tracked vehicle with a 7.5 cm Gebirgs-Geschuetz mounted at the rear ready to fire on a wheeled carriage with short trails. However, this proposal was not pursued. After various further proposals for guns mounted on other commercially available vehicles, by October 1927 Krupp had redirected

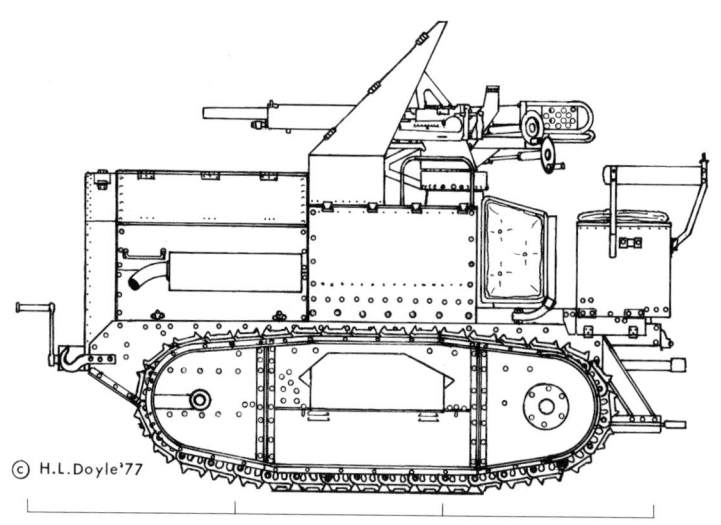

their efforts toward a self-propelled gun on a new and unique fully-tracked chassis design.

Rheinmetall-Borsig's involvement in the design of anti-tank guns mounted on commercially available chassis are revealed in an overview report from July 1943 on the weapons and military vehicles designed and developed for the Deutsche Wehrmacht since 1895. Within the 23 pages listing their numerous design projects for the army, navy, and air force were a 3.7 cm Selbstfahrlafette auf Hanomag-Schlepper and a 7.7 cm Kanone auf Hanomag-Schlepper created during the period from 1922 to 1932. An undated summary report from 1940 on Rheinmetall-Borsig artillery development from 1922 to 1940 included data sheets with photographs of a 3.7 cm WD-Schlepper 25 PS and a 7.7 cm WD-Schlepper 50 PS. In this report Rheinmetall stated that development of both projects was started in 1927 on orders from O.K.H. Ordinary commercially available model WD caterpillar tractors from Hanomag were outfitted with guns on pivotal mounts for employment as self-propelled guns. The 3.7 cm gun L/45 (45 calibers in length) fired a Pzgr. weighing 0.695 kg at a muzzle velocity of 760 m/s. Elevation arc was -10 to +15 degrees with a traverse arc of 30 degrees. The 7.7 cm gun fired a projectile weighing 6.8 kg at a muzzle velocity of 465 m/s. Elevation arc was -7 to +15 degrees with full 360 degrees traverse.

On 28 June 1928, the Wehramt reported on the status of the 5-year (1929-1932) procurement program for Kampfwagenabwehr (anti-tank defense): *The basis for continuing work on the procurement program is acquisition of 364 Tak (either horse drawn or pulled by a Schlepper), 138 Tak (mounted on a Schlepper), 34 Tak gun tubes for leicht Traktor (light tanks), and 40 Tak gun tubes for Pzkw. (armored cars). The costs associated with procuring pivot mounts and the work of converting 128 Schleppern to Selbstfahrlafetten must be determined. The 7.5 cm program continues as planned with 34 7.5 cm guns for Grosstraktor and 6 7.5 cm guns for experimental pieces for Kampfwagenabwehr.*

On 2 September 1929, General von Blomberg, head of the Truppenamt, wrote to the Wehramt on the subject of Tankabwehrzwischenloesung (anti-tank interim solution): *The Truppenamt has studied the final report sent by the Wehramt on 7 August 1929 on trials with a L.H.B. Schlepper 50 PS (full-tracked caterpillar tractor with 50 horsepower engine made by the Linke-Hofmann-Busch-Werke AG (later renamed Famo)). From this report it appears questionable if this Schlepper is suitable for mounting a 3.7 cm Tak. It appears to be very tall, and its width presents a large target. Also its cross-country mobility is insufficient. Therefore it is questionable if this design should be accepted into the army. The search for a usable Tankabwehrzwischenloesung appears to be urgent. Also on this subject, the Truppenamt would be thankful for a copy of the Wehramt report dated 1 May 1929 on their views on the W.D. Schlepper 28 und 50 PS.*

The Waffenamt reported on the results of the demonstration of motorized Tak-Zwischenloesung held in Kummersdorf on 28 January 1930: *Wa Prw presented a Linke-Hoffman Schlepper with Tak on a pivot mount, a Maffey Schlepper with Tak on a pivot mount, a Hanomag Schlepper towing a Tak, and a Marburger Traktor towing a Tak and ammunition trailer. The demonstration proved that none of these solutions meet basic requirements. Therefore, it was decided to stop trials of motorized Tak Zwischenloesung. The head of the Waffenamt agrees that a Tak-Jaeger needs to be designed.*

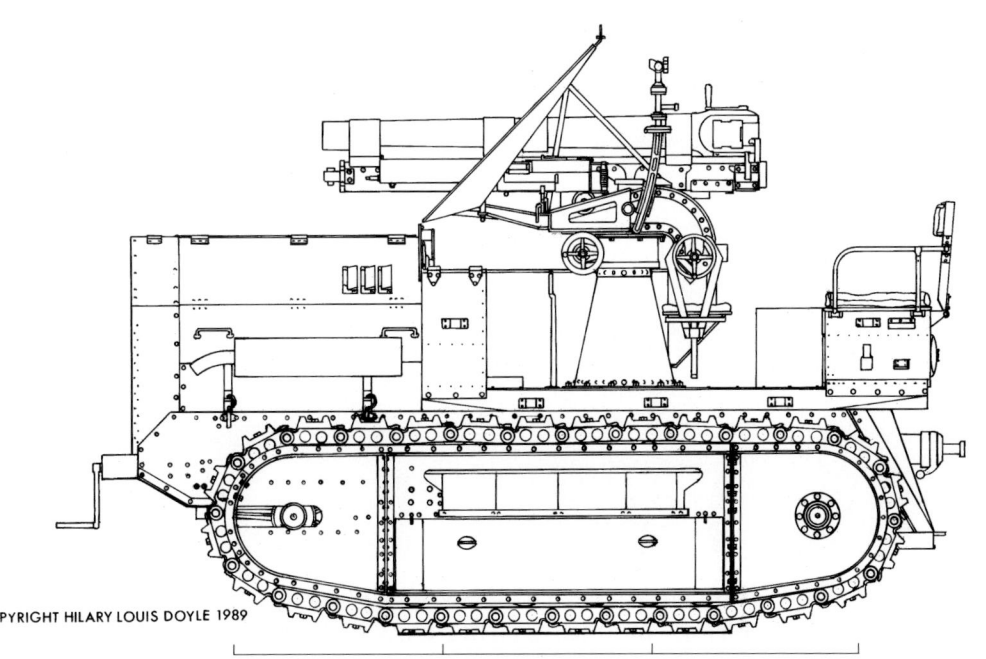

3,7 cm und 7,5 cm Geschuetz auf Motorlafetten (Tankzerstoerer) renamed leichte Selbstfahrkanone (L.S.K.) in 1930

By October 1927, the Artillerie-Konstruktion department at Krupp was working toward creating a full-tracked chassis specifically designed for a self-propelled gun. Initially known as a Motorlafette, the name was changed later to leichte Selbstfahrkanone, abbreviated as L.S.K.

Following a meeting on 12 November, Krupp sent Wa Pruef 4, sketch AKF 1558 dated 21 November 1927 of a 7.5 cm Infanterie-Geschuetz auf Motorlafette with preliminary design data and rough weight estimates. Details included:

As requested, the elevation has been increased to 70 degrees and the traverse arc to 30 degrees. The gun tube and breech are identical to the leichte Feldkanone P; however, the recoil has been shortened to 400 mm. Elevation must be reduced to about 30 degrees for the gun to be loaded. The recoil cylinder and recuperator are mounted under the gun to reduce the size of the penetration and lower the height of the gun shield. The upper carriage with short trunnions has a claw guide on the vehicle frame. The center of the gun has been moved about 50 mm to the left of the vehicle centerline to obtain additional space for the driver. At higher elevation, the carriage and gun tube recoils into an opening in the platform. There isn't any counterbalance. A weight is mounted forward on the gun tube and an additional weight on the lengthened carriage so that the center of gravity is at the shield trunnions. There are hand wheels on both the left and right side for elevation as well as capability to fire the gun from either side.

There is a crew of three men. The Richtkanonier (gunner) sits to the left of the gun and traverses with the carriage. He operates the traversing gear and then the elevating gear and firing mechanism on the left side. The Fahrer (driver) is seated on the right of the gun about 400 mm farther forward than the gunner due to the traversing arc of the gun. In the firing position the driver tilts the seat back and in a backward leaning position operates the breech and then the firing device and elevation gear. While driving, the Ladekanonier (loader) sits on the transmission casing and loads the rounds.

The frontal armor protecting the gun and gunner is still set back about 350 mm to the rear of the armor for the driver. It will be attempted to move the gunner forward so that single-piece frontal armor can be achieved and the opening for the gun tube and carriage can be adequately covered.

76 rounds of two-piece ammunition are stowed with two rounds per basket.

Statistics for the 7.5 cm Infanterie-Geschuetz

Above and Right: The first Motorlafette chassis designed by Krupp to meet the military specifications for a self-propelled carriage mounting either a 3.7 cm Tak or 7.5 cm Kanone. (WJS)

auf Motorlafette are: Projectile weight of 6.3 kg fired at a muzzle velocity of 465 m/s. Maximum range of 9600 meters with an elevation arc of -10 to +70 degrees, and total traverse arc of 30 degrees. Firing height above the crew platform is 550 mm and is 950 mm about the ground. Overall height of the Motorlafette with gun is 1.450 m, vehicle width 2.000 mm, track centers 1.600 m, vehicle length 3.850 m, and ground clearance 250 mm. The gun weighs 800 kg, 76 rounds of ammunition with packaging 825 kg, vehicle with 100 PS engine 4500 kg, frontal armor 250 kg, armor for engine, ammunition, etc. 900 kg, and crew of 3 men 225 kg for a total weight of 7500 kg.

On 25 November Krupp sent Wa Prw 4 new sketches AKF 1559, 1560 and 1562 of their preliminary design of 3,7 cm und 7,5 cm Geschuetz auf Motorlafetten (Tankzerstoerer) with the following explanation: *Because the prone position of the crew especially hindered the work and view of the driver, we have tried sitting positions in both of these conceptual designs. However, then the overall height can't be reduced below 1300 mm.*

As shown in sketch AKF 1462, with the crew in a prone position the overall height is about 1.000 m. However the gun sight and mirror for the driver extend 250 mm above this height and must be protected by a hinged upper shield in combat. The overall height with raised shield will then be about 1300 m - the same as for seated crew members. This is not such a disadvantage, since during combat the loader can work kneeling or bent over so that a higher rate of fire can be expected than when the men are prone.

Plans are to stow the required ammunition load of 30 rounds for the 7.5 cm and 100 for the 3.7 cm guns on the vehicle. With seated crew and 1.300 m height, the ammunition load increases to 76 rounds for the 7.5 cm and 150 rounds for the 3.7 cm.

Statistics for the 7,5 cm Geschuetz L/25 (Tankzerstoerer) auf Motorlafette (sketch AKF 1559 dated 24Nov27) are: Projectile weight of 6.3 kg fired at a muzzle velocity of 465 m/s. Maximum range of 6000 meters with an elevation arc of -7.5 to +15 degrees, and total traverse arc of 15 degrees. Firing height above the crew platform is 450 mm and is 850 mm above the ground. Overall height of the Motorlafette with gun is 1.300 m, vehicle width 2.000 mm, track centers 1.600 m, vehicle length 3.850 m, and ground clearance of 250 mm. The gun weighs 800 kg, 76 rounds of ammunition with packaging 825 kg, vehicle with 100 PS engine 4500 kg, 10 mm frontal armor 250 kg, armor for engine, ammunition, etc. 900 kg, and crew of 3 men 225 kg for a total weight of 7500 kg.

Statistics for the 3,7 cm Geschuetz (Tankzerstoerer) auf Motorlafette (sketch AKF 1560 dated 24Nov27) are: Projectile weight of 0.7 kg fired at a muzzle velocity of 800 m/s. Maximum range of 6500 meters with an elevation arc of -7.5 to +15 degrees, and total traverse arc of 15 degrees. Firing height above the crew platform is 450 mm and is 850 mm above the ground. Overall height of the Motorlafette with gun is 1.300 m, vehicle width 2.000 mm, track centers 1.600 m, vehicle length 3.850 m, and ground clearance of 250

mm. The gun weighs 500 kg, 150 rounds of ammunition in 15 bins 400 kg, vehicle with 100 PS engine 4500 kg, 10 mm frontal armor 250 kg, armor for engine, ammunition, etc. 900 kg, and crew of 3 men 225 kg for a total weight of 6800 kg.

A full-scale wooden model was assembled and shown by Krupp to the Waffenamt on 29 June 1928. During the demonstration, the Krupp representative described the design as follows:

The 7.5 cm weapon, which you see in front of you as a wooden model, is something new for the Artillerie of the deutschen Heeres (German army) in that it is not towed by draft animals or the crew. Instead it is sent against the enemy by motorized power. It is capable of indirect fire from hidden positions as well as direct fire at close combat ranges, to quickly exploit a breakthrough of enemy positions, and quickly and effectively hit swiftly moving ground targets.

The following basic specifications for the gun and vehicle result from these desired combat capabilities:

o Gun elevation of 45 degrees is needed for firing from hidden positions and achieving the longest range.
o Firing while moving is needed for following targets during a breakthrough.
o A large traverse arc is needed for firing at moving ground targets.
o Cross-country mobility is needed for all types of terrain.
o The lowest possible weight is needed for high speed on roads and cross-country.

This model was created to meet requirements established by the Waffenamt. This design ripened over a period of 18 months in which a series of conceptual designs were presented and discussed with the Waffenamt.

There were two possibilities for locating the gun and engine in the vehicle - engine at the front and motor rear - or vice versa. A front-mounted engine had the following disadvantages:

o To evenly distribute the weight, the gun must be mounted at the back end of the vehicle with the crew manning the gun standing on the ground resulting in a longer period to prepare the gun for action, a narrow traverse arc, and it isn't possible to fire while moving.
o As required by Wa Prw 6, the driver must have a good view of the ground ahead and therefore must sit in the front. Because he must remain in his position to turn the vehicle, the gun fires over him resulting in an unnecessary increase in firing height and overall vehicle height.
o In addition, a front-mounted engine had to be especially effectively protected with armor resulting in an increase in weight.

This and other disadvantages, such as unfavorable ammunition stowage, pushed us toward mounting the gun at the front and the engine in the rear resulting in good weight distribution and sufficient room for the crew and ammunition stowage.

After the basic question of mounting the gun on the vehicle was clarified, we began to fulfill the major requirements for minimizing the size, achieving the lowest possible height, and highest possible traverse arc. Wa Prw 6 had limited the overall weight to 6 metric tons in order to achieve 40 km/hr speed with an engine that wasn't very large. Wa Prw 4 also required the heaviest armor possible.

To achieve the lowest possible overall height, the gun was mounted at the front of the vehicle on the lowest platform possible. In this arrangement, the gun can be traversed through an arc of about 300 degrees, but not directly toward the rear where the housing for the commercially available engine is higher. This traverse arc of 300 degrees can be used only when the vehicle is stationary. It is limited to 40 degrees when the vehicle is moving because of the requirement for an unimpeded forward view for the driver, who can't be disturbed by the gun or gun crew.

It is still questionable whether the gun can be effectively fired on the move and if the normal rocking of a moving vehicle won't reduce accuracy to where every shot is merely a waste of ammunition. I suggest that on level ground, good roads, freshly plowed fields, or firm stubble fields, an experienced gunner can still hit targets while on the move.

The weight limit of 6 tons controls the degree of armor protection. The crew are well protected by a large 8-mm-thick shield mounted in front of the gun and the engine compartment has 6 mm thick armor. If Wa Prw 4 increased the weight limit, the sides could have 6 mm armor and a hinged 6-mm-thick side shields about 720 mm high could be added. These shields could be folded out and to the rear when firing the gun.

The resulting vehicle is intended for two different guns - an Infanteriegeschuetz with the capabilities of the leichte Feldkanone and a 3,7 cm Tankabwehrkanone with the capabilities of the towed Tak. The lower weight of the 3.7 cm allows heavier armor protection with a 10-mm-thick front shield and 8-mm-thick fore and aft side shields.

The measurements of the gun position in both width and height and therefore the overall height and width of the vehicle are governed by the space needed for the crew and the requirement for a large traverse arc. Measurements of the gun itself would allow a lower profile. However, in order to ensure vehicle mobility and achieve effective fire, the driver and gunner work in comfortable positions. That governs the overall height of 1650 mm above the ground and the overall width of the vehicle of 2300 mm including the tracks. An attempt to design a narrower vehicle was given up because the gun with crew couldn't traverse to the rear.

Above and Below: The unique design of the Motorlafette suspension in which the sprocket and idler pivoted with the roadwheels in an attempt to provide a better ride over obstacles. (WJS)

7-9

The resulting height of the vehicle was also determined by sufficient ground clearance. Considering the obstacles that had to be crossed, 300 mm ground clearance was selected.

The remaining space between the gun and engine for the crew and ammunition is tight but sufficient. Behind this is the armored engine compartment with the engine mounted transversely across the width of the vehicle. The height is caused by using a normal commercially available engine. The armored radiator is located behind the engine. An 8-speed transmission along with differential/brake steering gear is positioned in front of the engine. The brake assembly prevents the inside track from completely stopping in curves. The smallest inside steering radius is about 5 to 6 meters. When steering through curves, the driver brakes the tracks by turning a lever on the steering column.

The tracks are made of rubber. Steel links are inserted and held by the rubber and also fastened to two steel wire cables. The cables take the pull and the rubber permits noiseless running. The weight of the vehicle is distributed by 14 roadwheels onto the tracks, resulting in a very light specific load on each roadwheel with a consequently favorable degree of efficiency.

The drive sprockets for the tracks are not rigidly fixed to the hull as is the usual method, but are mounted on pivoting arms. The idler wheels are also mounted on pivoting arms. This completely original design, proposed by Wa Pruef 6, permits smooth passage over rough road surfaces and bumps in open country.

A further advantage of this vehicle is that the guide roadwheels are close to the road and raised up only when encountering an obstacle. By means of an hydraulic adjustment fitted at the back of each, the springs can be locked during firing. A rear towing pintle and two towing hooks in front are contemplated. Furthermore, on the left side of the vehicle a winch can be mounted for its own use or that of other vehicles in difficulty. The weight of the vehicle permits adequate fuel and oil supply for about 150 km.

This vehicle, designed for the guns of different caliber, can also serve as an ammunition carrier. All vehicles will be fitted with the base ring for mounting a gun, and provision has been made that in case of vehicle damage, the guns can be lifted by means of a simple light lifting tackle or in buildings by means of pulley and tackle for mounting onto another vehicle in running order.

If I may now elaborate on the details of the gun mounts, I would say that the weapons will of course have to be mounted on central pivot mounts, which by themselves allow effective and rapid fire against moving targets. Furthermore, the guns differ fundamentally from other field guns insofar as that the recoil distance has to be considerably shorter than usual due to the lack of space behind the guns. The recoil forces have to be absorbed by the entire vehicle. Because the German army has no experience regarding the behavior of tracked vehicles when firing, and because it was therefore deemed prudent to keep the recoil forces as low as possible to reduce wear on the tracks, we were forced, especially in the case of the 7.5 cm gun, to make a compromise which we trust will be correct. Final judgment will depend on the outcome of trials.

The aiming speeds for traverse and elevation will be selected to enable quickly moving ground targets to be followed with ease. The elevation range of the 7.5 cm is -10 to +45 degrees and of the 3.7 cm -10 to +20 degrees. The sighting mechanism is so arranged that it is suitable for direct and indirect fire in the case of the 7.5 cm gun and direct fire for the 3.7 cm. We have designed a simple loading device for the 3.7 cm by which we hope to raise the maximum rate of fire to 50 rounds per minute. A crew of four has been envisaged; driver, gunner, loader, and gun commander. 80 rounds can be stowed for the 7.5 cm Kanone and 320 rounds for the 3.7 cm Tak.

Achievable ranges are:
9500 m for the 7.5 cm at 45 degrees elevation
7200 m for the 3.7 cm at 20 degrees elevation
3000 m for the 3.7 cm at 3 degrees 7 minutes elevation

Two meter high targets lie in the <u>Bestrichenen</u> (danger) zone at up 450 meters range from the 7.5 cm Kanone and 750 meters from the 3.7 cm Tak.

As recorded in their fiscal year report for 1928/29 (covering the period from 1 October 1928 to 30 September 1929) Krupp was awarded a contract worth 132,000 RM for the production of two Selbstfahrlafetten and spent 10,500 RM on their design.

In January 1930, Wa Pruef 6 also reported that the Horch company was developing a wheel-cum-track that was also intended to be utilized as a leichter Selbstfahrlafette fuer Tak (3.7 cm Kan und 7.5 cm l.F.K.). No other information has been found on this Horch project.

In the next fiscal year 1929/30 report, Krupp reported that two L.S.K. had been produced, test-driven at Meppen, and delivered to the army at a cost of 169,781 RM. As described in an original Krupp proposal for the L.S.K.:

This Motorlafette is a full-tracked vehicle. The Maybach O.S.5 engine rated at 100 metric horsepower allows the vehicle to climb 33% slopes and achieve speeds up to 40 km/hr. The vehicle can ford streams up to 98 cm deep.

The hull serves to mount major components including suspension components like the drive sprocket and idler wheels, roadwheels, and return rollers that the tracks run on. The hull is made out of 4, 6, and 14 mm thick carbon steel plates (not hardened ar-

mor) that are welded together and only bolted where assembly of individual components is needed. The thicknesses of the most important parts are 14 mm front wall, 6 mm side and rear walls, 6 mm roof, and 4 mm belly. The hull is strengthened by a transverse wall and riveted U-shaped ribs below the gun platform. The external side walls are removable for access to the suspension and have a large opening to allow mud to fall out. Two covered openings in the external side walls allow access to the track tensioners at the front and for installing the spill drum at the rear. Four covered ports in the side walls are for operating the dampening cylinders for the clamps. Two tow hooks are located at the front and a towing pintle is mounted on the rear of the hull.

The ground clearance is 300 mm. The gun is mounted forward and the engine at the rear in order to achieve a low overall height and a large traverse arc, as well as being able to fire on the move. The engine, transmission, rear-mounted radiator, and fuel tank are all mounted transversely across the hull in an enclosed engine compartment. The area between the gun and engine compartment is utilized for the crew and ammunition stowage.

The engine, transmission, and steering unit are mounted on a platform so that they can be pulled out together as a unit and, if necessary, exchanged for another drive train unit. The transmission is mounted parallel to the engine and attached by a transfer case with a 1:1 gear ratio. The transmission has four internal gears with ratios at 1:1, 1:1.48, 1:2.5, and 1:4 with an external auxiliary gear box providing two ranges - a 1:1 ratio for higher speeds on roads and a 1-3.36 ratio for lower speeds but higher torque for cross-country travel. Power is transferred through the steering unit to drive shafts penetrating the hull sides to the swing arms for the sprocket drive wheels. The maximum gear reduction from the engine to the drive sprockets is 1:105.

The engine compartment is decked over. Gratings at the upper rear of the engine compartment let in cooling air. The heated air is discharged through covered gratings on the roof. The deck over the transmission and steering unit is removable, and there are additional air slits in the firewall. Cooling air is pulled through the radiator mounted behind the engine and pushed out through gratings in the roof and slits in the firewall. Two openings in the belly below the fore and aft engine sumps are for draining the oil and in addition there is a threaded drain in the belly for draining the engine coolant.

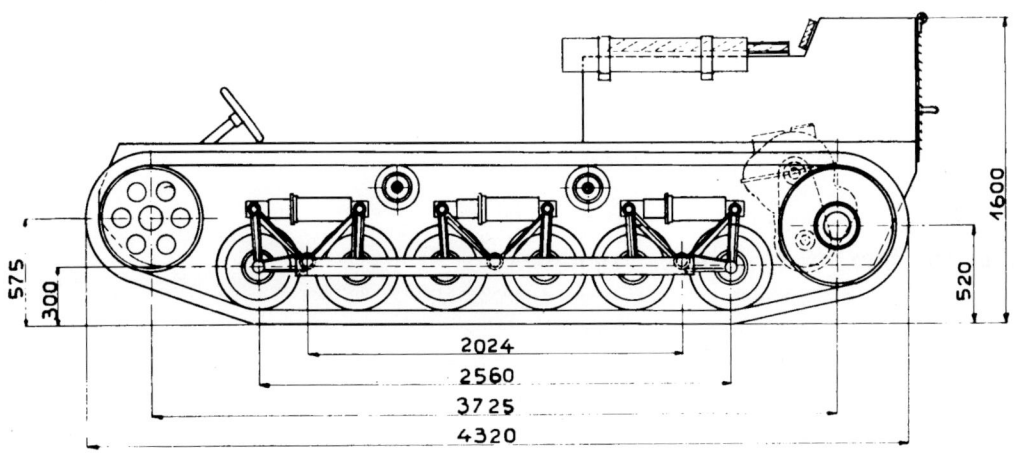

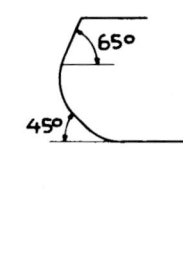

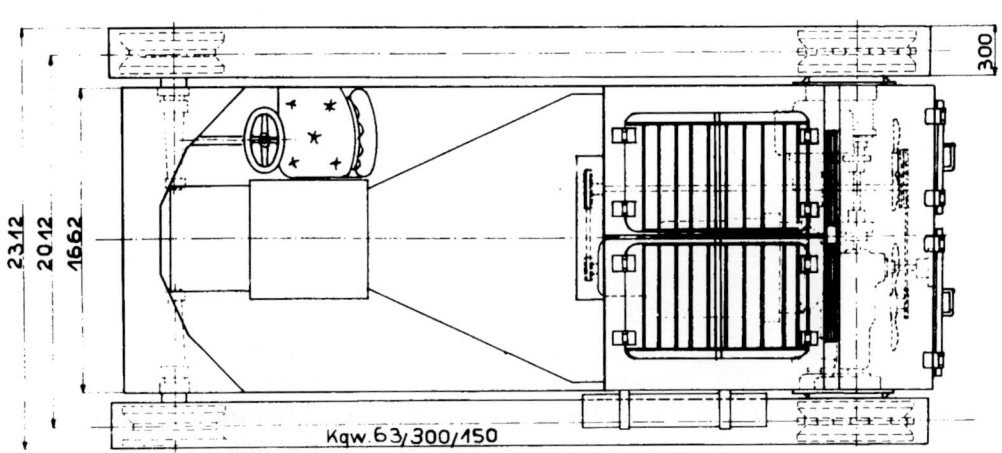

This Page:
The Krupp leichte Selbstfahrkanone (L.S.K.) after undergoing major suspension conversion to a simpler design with three pairs of roadwheels. It was intentionally designed to create the lowest possible height profile for a self-propelled gun carriage. (WJS)

The idler wheels, mounted on forked swing arms adjustable for track tensioning, are mounted high enough to climb over 60 cm high vertical obstacles. The swing arm mounts for the drive sprockets and idler wheels allow a smooth passage over rough terrain. When driving on good roads these swing arms can be blocked.

Three return rollers are mounted on each side. The vehicle's weight is distributed over 14 roadwheels that transfer the load to the tracks. With a track contact length of 2.6 meters, 1.5 meter wide trenches can be crossed. The 14 roadwheels are mounted in two groups, with each roadwheel arm acting against a common spring. Two clamps are installed to secure the spring when firing.

The vehicle has a maximum length of 4.200 meters, maximum width of 2.280 meters, maximum height (without the gun) of 1.580 meters. The chassis weighs 5,300 kg, the gun with shield 1.530 kg, ammunition 750 kg, and crew 300 kg for a total weight of 7,880 kg.

An L.S.K. was delivered to the Kraftfahrversuchsstelle Kummersdorf, arriving on 21 August 1930. The completely unique suspension design with pivoting drive sprocket arms did not work. After being driven for only 84 kilometers, the L.S.K. was returned to Krupp in November 1930 for installation of a new suspension. After designing and installing a completely different suspension consisting of six independently sprung roadwheels per side, the L.S.K. was returned to Kummersdorf in November 1931, where it was driven 379 kilometers in 1931, 1007 in 1932, and 589 in 1933.

After it was rebuilt, the appearance of the L.S.K. had totally changed from the original design. As reported in September 1932, the modified L.S.K. weighed a total of 8,910 kg. There were 86 links with 115 mm pitch in each track. The six roadwheels per side resulted in a track contact length of 2.560 meters.

Not knowing that the letter "k" in L.S.K. stood for Kanone, by 1933 it had become known as the L.S.Krupp at Kummersdorf. Statistics from a 1933 report reveal that the L.S.K. still was powered by a 6-cylinder Maybach O.S.5 engine, rated at 100 metric horsepower at 1900 rpm, with a four-speed transmission, and a Cletrac steering unit. Overall length was about 4.500 meters, width at 2.515 meters, height without gun of 1.635 meters. The assembled L.S.K. with gun, crew, ammunition, and 195 liters of fuel weighed a total of 9,100 kg. Mobility characteristics were listed as 30 cm ground clearance, 33 degree slope, 55 cm step, 96 cm ford, and 1.5 m wide trench. The carbon steel hull (not hardened armor plate) was welded together with a 14-mm-thick front wall, 12 mm side walls, 10 mm rear wall, 6 mm deck, and 5 mm belly.

The 7.5 cm gun was developed and assembled under a separate contract. A summary report on artillery development written by a Krupp engineer in 1945, stated that: *The gun mount design had been completed by March 1930. The first test firing of the Mittelpivotlafette (pedestal mount) took place in November 1930, with the acceptance test firing occurring in January 1931. The gun proved a success with no specific problems. In July 1932 an electromagnetic firing mechanism was installed that was operated from the gunner's handwheel. This supplemented the single foot pedal trigger that was retained. At this same time, because of tight crew space, a combination safety switch was installed for the loader. This interrupted the firing mechanism after every shot (regardless of whether the gun was fired by the foot pedal or the electromagnetic trigger) until the safety switch was reset by the loader.*

The original intention of mounting a 3.7 cm Tak L/45 on the L.S.K. was never realized. Waffenamt lost interest in this leichte Selbstfahrkanone, and the development program was dropped.

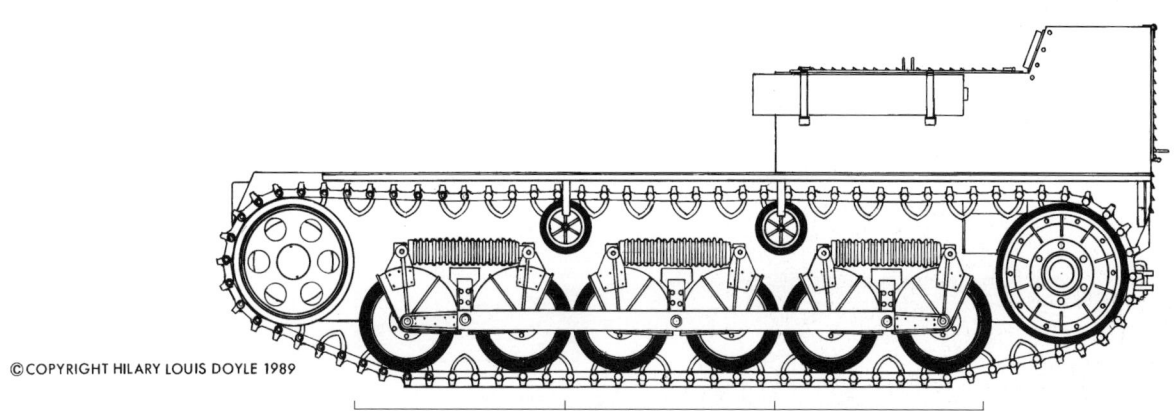

Leichte Selbstfahrkanone modified by Krupp in 1931

Leichttraktor Rheinmetall Selbstfahrlafette

On 19 July 1928, Inspektorat 6 sent the tactical specifications for the design of the Leichttraktor to the Waffenamt with the request that Wa Prw award contracts to Daimler, Krupp, and Rheinmetall for the design and production of two Leichttraktor each. If Daimler didn't accept the offer, then Wa Prw was requested to give a contract to one of the remaining two firms for a third chassis as a gepanzerter Nachschubfahrzeug (armored supply vehicle) and a contract to the other firm for a third chassis as a 3.7 cm Kampfwagen-Abwehr-Selbstfahrlafette.

Based on a meeting with Wa Prw 6 on 26 May 1928, Krupp's Abteilung Artillerie-Konstruktion had started on the design of the L.Tr. as a self-propelled gun. On 3 July 1928, Krupp reported to Wa Prw 6 that: *A proposal for the L.Tr. als Selbstfahrer fuer 3.7 cm Kanone is being prepared. However, due to its current dimensions, it is not likely that the L.Tr. can be used as a Selbstfahrer fuer 7.5 cm Feldkanone because of the lack of space for the crew and ammunition.*

The contract for the detailed design and production of a single 3.7 cm Selbstfahrlafette was awarded to Rheinmetall instead of Krupp. Rheinmetall completed the chassis, mounted the 3.7 cm gun in a turret, and delivered the L.Tr.Rheinmetall Selbstfahrlafette to the Kraftfahrversuchsstelle Kummersdorf in October 1930. At Kummersdorf, the L.Tr.Rhm.Sfl. was driven 624 km in 1930, 123 in 1931, 223 in 1932, and 307 in 1933.

Initially the L.Tr.Rhm.Sfl. had the same suspension as a normal L.Tr. from Rheinmetall with 12 small roadwheels. This suspension was subsequently replaced with an entirely new design that featured a suspension with four pairs of roadwheels, each pair sharing a large coil spring. The Kgs.61/301/90 cast steel tracks for the modified suspension had a center guide horn.

Statistics from an August 1933 report reveal that the modified L.Tr.Rhm.Selbstfahrlafette was powered by a 6-cylinder, 7.8 liter, Mercedes-Benz M 36 engine, rated at 100 metric horsepower. It had a synchronized 4-speed G-55 transmission and a Cletrac clutch/brake steering unit. Maximum speed was 30 km/hr with a range of about 137 km (based on fuel tank capacity of 220 liters) with 160 liters/100 km fuel consumption. Overall length was about 4.340 meters, width at 2.106 meters, height without gun of 1.450 meters. The total weight of the L.Tr.Rhm.Sfl. was 8,735 kg, with the chassis alone weighing 6,735 kg. The carbon steel hull (not hardened armor plate) was made up of plates from 5 to 14 mm thick. A crew of four men drove the vehicle and manned the 3.7 cm Kanone. 150 rounds were stowed for the 3.7 cm Kanone along with 3000 rounds for a machinegun.

The L.Tr.Rhm.Sfl. was identified in the Rheimetall-Borsig overview report from July 1943 as their 3.7 cm Selbstfahrlafette L/45 (3.7 cm Pak L/45). A 1940 summary report on Rheinmetall-Borsig artillery development from 1922 to 1940 included a data sheet on the 3.7 cm Selbstfahrlafette L/45 with a photograph showing the original production version from 1930 prior to modification. In this report Rheinmetall stated that development of this project started in 1934 [sic] on orders from O.K.H. [sic]. A new turret housing the 3.7 cm Pak L/45 with a light machinegun was produced for the Rheinmetall Leicht-Traktor. Statistics from this report stated that the 3.7 cm Pak L/45 fired a Pzgr. weighing 0.695 kg at a muzzle velocity of 760 m/s out to a maximum range of 6,400 meters. Elevation arc was -10 to +25 degrees with all-round traverse of 360 degrees. The engine rated at 100 metric horsepower propelled the 8.5 metric ton vehicle at speeds up to 35 km/hr. It was manned by a crew of 3 [sic] and had 13-mm-thick armor [sic] protection.

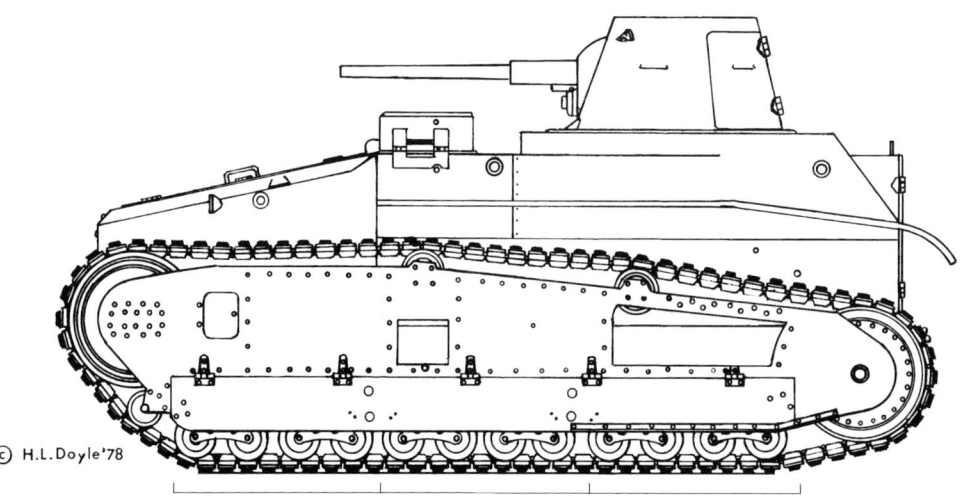

7-15

7-16

This Page:

The Leichttraktor Selbstfahrlafette with 3.7 cm Tak in a fully traversable turret as originally assembled by Rheinmetall in 1930 directly before being sent to Kummersdorf for trials. (WJS)

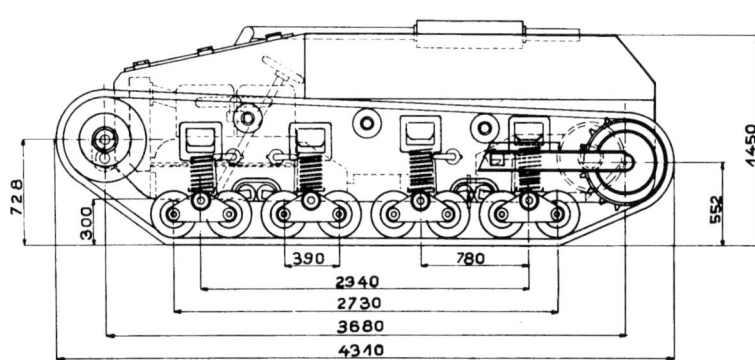
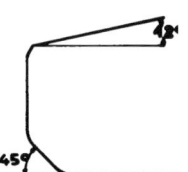
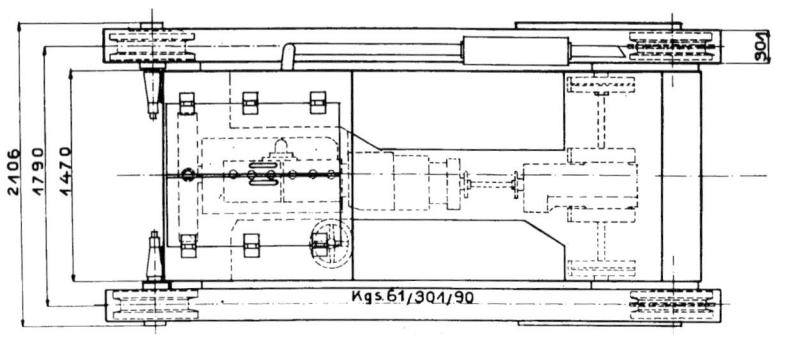

This Page:
The Rheinmetall Leichttraktor Selbstfahrlafette in 1933 after having undergone extensive modification, with four pairs of larger diameter roadwheels (each pair sharing a large coil spring) replacing the original suspension.
(WJS)

3.7 cm und 7.5 cm Selbstfahrlafette auf Halbketten-Fahrgestell

A program had been initiated by 20 November 1934 to create Panzerjaeger that were much faster and more maneuverable than tanks. Two different vehicles were identified in April 1935 that were to be tested at Wunsdorf starting in the Summer of 1935 and ending in the Spring of 1936. These were a 3.7 cm Panzerabwehr-Geschuetz auf Selbstfahrlafette and a "Tankjaeger".

Further details about the specifications and characteristics of these two highly mobile anti-tank weapon systems were revealed in a report from Wa Prw dated 30 October 1935, entitled <u>Offensive Abwehr von Panzerwagen</u> (offensive defense against tanks):

During the demonstration at Kummersdorf on 11 July 1935, the Oberkommando des Heeres himself emphasized that a offensive tank defense must be strived for. The following statements about technical development have resulted from thoroughly working through this problem.

The frequently expressed principle that the best defense against a tank is another tank is at least disputed. Bomber aircraft are not employed against heavy bomber aircraft. Instead, faster and more maneuverable fighter aircraft presenting a smaller target are used that only have to be armed in such a way that they can really damage a bomber. At sea the torpedo boat is fought with a torpedo boat destroyer, the submarine with a submarine destroyer. In both types of destroyers the most important specification is superior speed and maneuverability -- armored only as far as it doesn't interfere with both primary characteristics. Recently this train of thought by many sailors has reached its peak in the Schnellboot. Apparently even against the heaviest and largest warships, the Schnellboot is a valuable and dangerous offensive weapon. These very small targets have a considerable speed of about 40 knots and therefore are extremely difficult to hit. Armor is practically renounced. They have fulfilled their role if they can get within effective torpedo firing range undamaged. Even if they are destroyed after achieving their task, their employment is worthwhile, even from a purely economic standpoint. But after firing torpedoes, they will frequently succeed in getting away, thanks to the characteristics of high speed, maneuverability, and small target area.

The question remains whether a tank destroyer can be built exactly corresponding to the fighter aircraft, torpedo boat destroyer, submarine destroyer, and especially the Schnellboot, in which armor is rejected in favor of the following primary characteristics:
1. very high road speed to achieve good operational mobility;
2. good cross-country mobility, at least the same as tanks;
3. cross-country speed and maneuverability greatly superior to a tank;
4. small target area;
5. good firing platform for accurate shooting while stationary;
6. weapons capable of target destruction at ranges from 700 meters up;
7. cheap and quick mass production in comparison to tanks;
8. if possibly achievable without interfering with the first seven requirements, rapidly dismountable weapon for the purpose of employment in the same way as towed anti-tank guns, as well as rapidly remountable.

This creates an entirely new automotive problem as well as a totally new weapons problem. The automotive problem on land is certainly very much more difficult than on the sea for a Schnellboot. Whether the problem can really be satisfactorily and economically solved is completely undetermined and requires thorough investigation. Some people are thoroughly skeptical. Still, these ideal requirements were created primarily in order to test the level at which available or newly designed vehicles can meet these specifications.

The weapons problem is especially difficult because the requirement for sufficient armor-penetration capability must be coupled with the requirement for the least possible recoil. In addition to a highly effective muzzle brake, the extent to which other means can be found to solve the problem must still be clarified. One could also think of employing the rocket effect, which makes a recoilless mount possible by firing large explosive charges instead of armor-penetrating rounds.

The question of how well current designs under development meet the above requirements is answered as follows:

<u>3.7 cm Tak auf Selbstfahrlafette (kleine gelaendegaengige Zugkraftwagen)</u>

In recognition of the disadvantages of the 3.7 cm Tak L/45 that has been procured for the infantry and the anti-tank battalions for divisions, a long time ago Wa Prw initiated the development of a Tak auf Selbstfahrlafette. The weapon selected was the standard 3.7 cm Tak L/45 accepted by the army which was to be mounted on a kleine gelaendegaengige Zugkraftwagen Fahrgestell with the engine mounted in the rear. Its maximum speed on roads was to be between 60 and 70 km/hr, varying according to the cross-country terrain but expected to be 25 to 30 km/hr across plowed fields. The high road speed (more than double that of a tank) allows anti-tank units to be rapidly sent to threatened sectors. Its cross-country speed and mobility are also superior to that of tanks. However, the requirements for an ideal Tankjaeger are only partially fulfilled with this design. In comparison to these ideal requirements (but not in comparison to other tank designs), it still has several disadvantages, including a high profile and the Tak can't be readily dismounted from the Selbstfahrlafette - at least not from the first trial vehicle.

Armor protection is the same as leichten Pan-

Left and Below:
**This 3.7 cm Selbstfahrlafette L/70, designed and produced by Rheinmetall on the Hansa-Loyd HL kl 3(H) chassis with rear mounted engine, was photographed on 28 September 1936. This type of high-speed <u>Tankjaeger</u> was intended to be the army equivalent to a fighter aircraft for an air force or a torpedo boat for a navy.
(RC)**

zerspaehwagen, also open on top and capable of stopping penetration by S.m.K. rounds. The first trial vehicle is to be completed in the Spring of 1936.

In recognition of the increased armor protection on French tanks, mounting a higher performance 3.7 cm Tak L/65 onto the kl.gl.Zgkw. has been examined. This longer gun is to have a muzzle brake and a high muzzle velocity of 880 m/sec. This L/65 gun can be mounted on the kl.gl.Zgkw. at any time. However, switching to this gun with increased penetrating capability means that different ammunition is needed than that used in the standard 3.7 cm Tak L/45.

7.5 cm Kanone auf Selbstfahrlafette (leichte gelaendegaengige Zugkraftwagen)

With regard to the possibility of defending against the heaviest French tanks (Char 2 C, 3 C, and D), a 7.5 cm Kanone auf Selbstfahrlafette was developed on orders from the Oberkommando des Heeres himself. The gun has the capabilities of the Feldkanone 15 n.A., achieving a muzzle velocity of 650 m/sec. Fired at this initial velocity, at a range of 700 meters the Panzergranate can penetrate 61 mm thick armor plate hit at an angle of 30 degrees from vertical. This is 10 percent higher than what is needed to penetrate the Char 2 C, 3 C and 2 D.

The leichte gelaendegaengige Zugkraftwagen, serving as the chassis, has a maximum road speed of about 70 km/hr, good cross-country mobility, and about the same cross-country speed as tanks. The high road speed (more than double that of tanks) allows this self-propelled gun to be rapidly switched to threatened sectors. But all of the requirements for an ideal Tankjaeger have not been fulfilled in this design. The first trial vehicle is to be completed in the Spring of 1936.

3.7 cm Selbstfahrlafette L/70

Wa Prw awarded contracts to Rheinmetall-Borsig for detailed design of the weapon and turret and Hansa-Loyd for the chassis. The 3,7 cm Tak auf Sfl. (kl.gl.Zgkw.) was identified in the Rheimetall-Borsig overview report from July 1943 as their 3.7 cm Selbstfahrlafette L/70 created during the period from 1933 to 1939. A 1940 summary report on Rheinmetall-Borsig artillery development from 1922 to 1940 included a data sheet on the 3.7 cm Selbstfahrlafette L/45 with a photograph taken in 1936. In this report Rheinmetall stated that development of this project started in 1935 on orders from O.K.H. A turret for the 3.7 cm Pak L/70 and two light machineguns designed by Rheinmetall was supported on a pedestal mount. Statistics from this report stated that the 3.7 cm Pak L/70 fired a Pzgr. weighing 0.710 kg at a muzzle velocity of 900 m/s. Elevation arc was -7 to +20 degrees with all-round traverse of 360 degrees.

From data gathered in the postwar BAOR Technical Intelligence Report from June 1946: Hansa-Loyd designed and produced a few model HL kl 3(H) semi-tracked chassis with the engine mounted in the rear. The HL kl 3(H) was provided with armored bodywork and carried a 3.7 cm Pak L/70 and two light machine-guns in the turret.

Power was provided by a 3.485 liter, 6-cylinder Borgward engine rated at 70 metric horsepower at 2600 rpm. It had a 4-speed Zahnradfabrik manual transmission with 2-speed transfer case. The five roadwheels per side with torsion bar suspension were spaced to provide a track contact length of 1.600 m. The overall length was 5.100 m, width 2.000 m, front wheel base 1.650 m, and track base 1.600 m. Its maximum road speed was 50 km/hr.

In their 1940 summary report, Rheinmetall-Borsig stated that only one trial 3.7 cm Sfl. L/70 had been completed.

7,5 cm Selbstfahrlafette L/40,8 (Modell 1)

Wa Prw awarded contracts to Buessing-NAG for the detailed design and assembly of the chassis and to Rheinmetall-Borsig for the gun and superstructure. This 7.5 cm K. auf Sfl. (le. gl.Zgkw.) was identified in the Rheimetall-Borsig overview report from July 1943 as their 7.5 cm Selbstfahrlafette L/40.8 Model 1 and 2 created during the period from 1933 to 1939. A summary report on Rheinmetall-Borsig artillery development from 1922 to 1940 included a data sheet on the 7.5 cm Selbstfahrlafette L/40.8 (Modell 1) with a photograph taken in 1940. In this report Rheinmetall stated that development of this project started in 1934. Statistics from this report stated that the 7.5 cm Geschuetz L/40.8 fired a Pzgr. weighing 6.8 kg at a muzzle velocity of 685 m/s. Elevation arc was -9 to +20 degrees with all-round traverse of 360 degrees. Armor plates on the front and sides were 20 mm thick and 8 mm thick for the deck and belly. It was manned by a crew of 4.

From data in a postwar BAOR Technical Intelligence Report from June 1946, a few chassis of an experimental model BNL6(H) were produced. These were intended as armored vehicles and mounted a 7.5 cm L/40.8 gun in a turret. The vehicles had five roadwheels per side and the engine was mounted at the rear. Buessing-NAG also developed another rear-engine model, the BN10(H), rather resembling the BNL6(H), and mounting the same gun. A Buessing NAG manual dated 1943 lists the Fgst.Nr. series as 2006 to 2008 for the BN10(H) chassis.

Surviving photographs and drawings help fill in a few additional details. A full-scale wooden model was completed by 1 August 1938. Except for the number of roadwheels, this wooden model resembles the 7.5 cm Sfl. L/40.8 with a model BN6(H) chassis that was photographed on 5 August 1938.

This and Opposite Page:

Photographed in August 1938, this 7.5 cm Selbstfahrlafette L/40.8 (Modell 1) was assembled by Rheinmetall on a Buessing-NAG BNL6(H) chassis with the engine mounted in the rear. (RC)

7978
5.8.38

7-23

This and Opposite Page:

Photographed in January 1940, this 7.5 cm Selbstfahrlafette L/40.8, Versuchsstueck 3 (Modell 1) was assembled by Rheinmetall on a Buessing-NAG BN10(H) chassis with the engine mounted in the rear.

The oval-shaped tube protruding out of the turret rear was part of the Huelsenauswerfereinrichtung (spent cartridge ejection mechanism).
(RC)

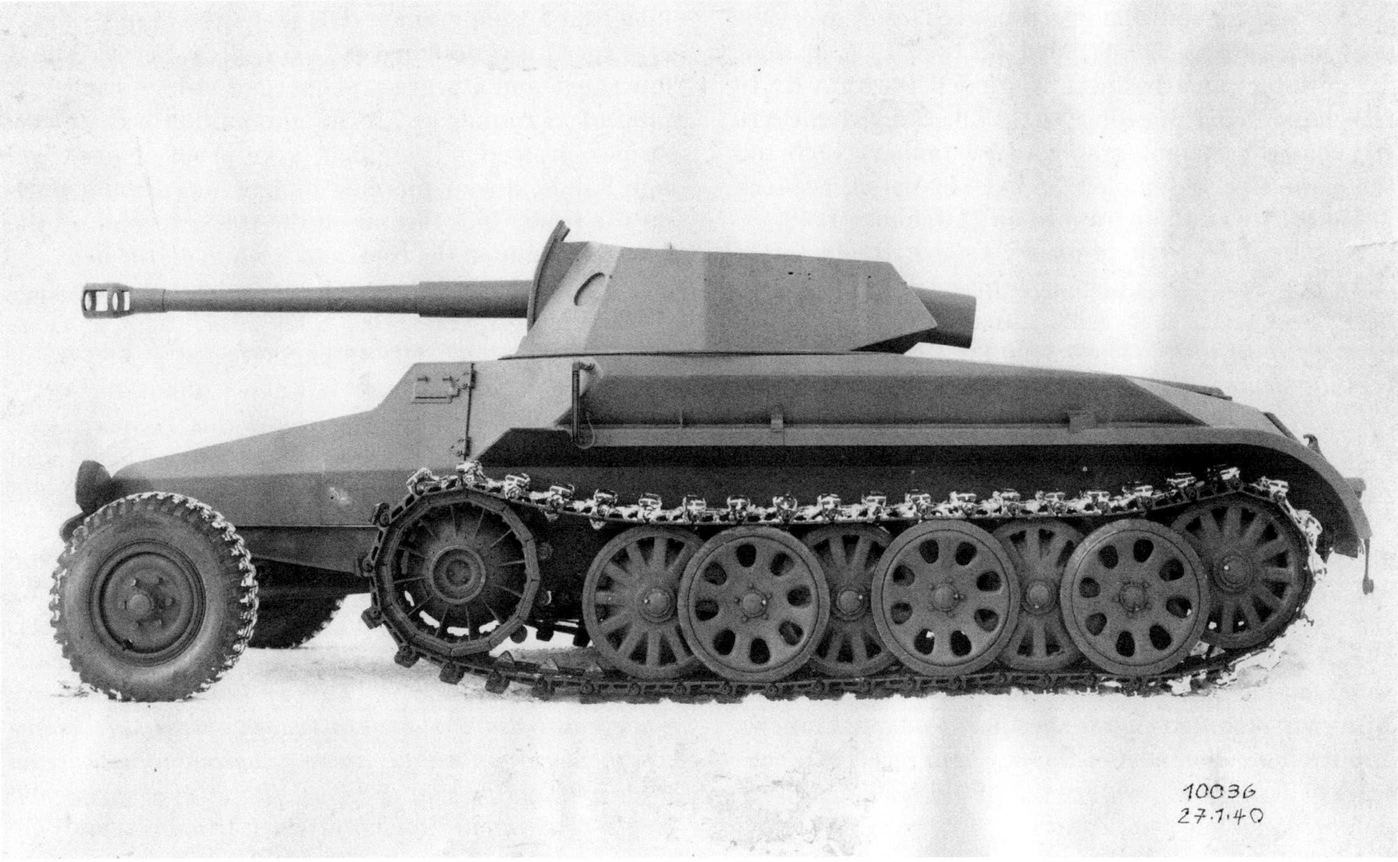
10036
27.1.40

WH
058057
10042
27.1.40

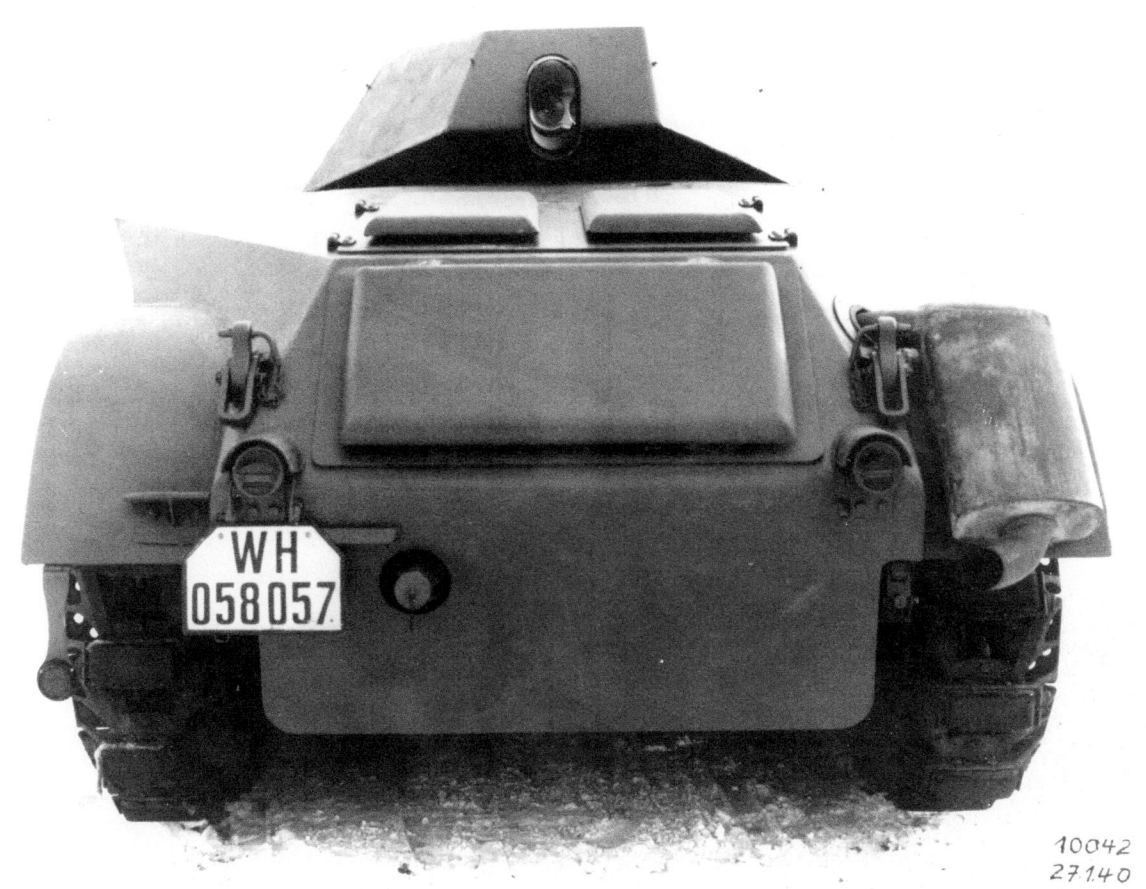

Rheinmetall-Borsig produced an overview drawing H-SK-A 51938 dated 17 June 1939 of their 7.5 cm Sfl. Versuchsstueck 3, Modell 1 with a BN10 (H) chassis from Buessing NAG. The completed BN10 (H) chassis was photographed on 9 January 1940 and the completely assembled 7.5 cm Sfl. Versuchsstueck 3, Modell 1 was photographed on 21 January 1940.

In their 1940 summary report, Rheinmetall-Borsig claimed that altogether three trial models had been produced, each with different firing heights, lengths, and widths. The 7.5 cm Sfl. L/40.8 (Modell 1) was still being tested at the time of this report in 1940.

Panzer-Selbstfahrlafette II - 7.5 cm Kanone L/41 auf Zugkraftwagen 5t (HKP 902)

As revealed in a Waffenamt summary report dated 1 July 1942, a new design for a Panzerjaeger auf Selbstfahrlafette was initiated in response to the request of the General Stab des Heeres in March 1936. A contract was given to Buessing-NAG to develop an advanced generation half-tracked chassis with rear-mounted engine specifically adapted for this application. Rheinmetall-Borsig again received a contract to design and produce the 7.5 cm Kanone L/40.8, mounted in a low profile turret.

The 7.5 cm Kanone L/40.8, mounted in a fully rotating open-top turret with elevation from -8 to +20 degrees, fired a 6.8 kilogram K.Gr.rot Pz. (APCBC-HE shell) at a muzzle velocity of 685 m/s and a 5.85 kilogram Sprenggranate (HE shell) at 485 m/s. Traverse speed was designed to be 3 degrees per second at low range and 8 degrees per second at high range. A total of 35 rounds of 7.5 cm ammunition was stowed. Armor protection, designed to be proof against 7.92 mm S.m.K. armor-piercing bullets, was 20 mm thick on the front, 14.5/10 mm on the sides, 10 mm on the rear, 10.5 mm on the roof, and 5.5 mm on the belly.

The HKP 902 chassis, assembled by Buessing-NAG in Fgst.Nr. series 2009-2012, had a rear-mounted Maybach HL 45 engine rated at 150 hp at 3800 rpm. Nothing else is known about the drive train that was designed to propel this 11 metric ton vehicle at a maximum speed of 50 km/hr. The crew of four consisted of a commander, gunner, loader, and driver.

In 1941, Rheinmetall-Borsig completed the assembly of two trial Pz.Sfl.II - 7.5 cm Kanone L/41 auf Zgkw.5t (HKP 902) in armor. On 11 August 1941, Organisations-Abteilung (III) ordered: *The two Panzer-Jaeger (Sfl.) - 7.5 cm Kanone L/41 auf gepanzerte Fahrgestell des 5 t Zugkraftwagen, which are either available or soon will be, are to be used to rapidly create a platoon and be sent to Libya. Because of their good armor-penetration capability, which remains almost the same out to a range of 1500 meters, and their maneuverability, these guns were also to be employed as light Schartenknacker (bunker busters) against fortifications.*

On 26 August 1941, detailed orders were is-

sued to create a schwere Panzer-Jaeger Zug (Sfl.) with 7.5 cm Pz.Sfl.II for the Afrika Korps at the Panzertruppenschule in Wunsdorf. This Zug was assigned as an integral part of Panzer-Jaeger-Abteilung 605. The first Pz.Sfl.II was to be prepared for tropical employment by 5 September 1941. The Waffenamt was to send the second Pz.Sfl.II to the Panzertruppenschule after assembly was completed.

This schwere Panzer-Jaeger Zug (Sfl.) consisted of 1 Zugfuehrer (platoon leader) with a machinepistol, 2 Geschuetzfuehrer (gun commander) with pistols, 2 Ladeschuetze (loaders) with pistols, 2 Richtschuetze (gunners) with pistols, 4 Kraftwagenfahrer fuer Sfl. (drivers, of which two were replacements) with pistols, 2 Panzerwarte (mechanics) with pistols, 3 Kraftwagenfahrer (2 car and 1 truck drivers) with rifles, 1 Melder (messenger) with rifle, 2.Kraftwagenfahrer (second driver) with rifle, 2 Pz.Sfl.II, 2 l.gl.Pkw.(Kfz.1), (Volkswagen with four-wheel drive), and 1 m.gl.Lkw. (open-bed truck for ammunition and equipment).

The initial ammunition issue consisted of 75 rounds of 7.5 cm Panzergranate and 35 rounds of 7.5 cm Sprenggranate for each gun.

As can be seen in the following reports, there was no fixed nomenclature used in describing the Pz.Sfl.II when the schwere Panzer-Jaeger Zug (Sfl.) was transferred to Libya:

7Dec41 - 1 7.5 cm Pz.Jg.Sfl. is moving from Germany to Italy.
5Jan42 - 1 7.5 cm Pz.Jg. arrived by ship in Tripoli.
15Jan42 - Supply officer of the 90.leichte Afrika Division reported that a 7.5 cm Sturmgeschuetz auf Sfl. for Panzer-Jaeger-Abteilung 605 had arrived in Africa and was on the way forward.
17Jan42 - Panzer-Jaeger-Abteilung 605 reported the arrival of a 7.5 cm Sfl.
23Feb42 - 1 gun for s.Sfl.Kdo.7.5 arrived in Tripoli on the ship Lerici.

The first Pz.Sfl.II had arrived in time to take part in Rommel's counterstrike on 21 January 1942. On 8 March 1942, the 7.5 cm Pak (Sfl.) Zug in the 90.leichte Division was transferred to the Kampfstaffel des Oberbefehlshaber Panzerarmee Afrika in exchange for three 7.62 cm Sfl.

Both of the 5 ton Sfl. 7.5 cm were reported present in an inventory list on 31 March 1942. But only one 7.5 cm schwere Panzer-Jaeger was reported operational with the Kampfstaffel PzAOK Afrika on 25 May 1942 at the start of Operation Venezia. No reports have been found on the fate of the second gun -- just undated photographs taken by the British after its capture.

One 7.5 cm Panzer-Jaeger was reported operational with the Kampfstaffel each day from 30 May to 2 June 1942. In their operational report for 5 June 1942, the Kampfstaffel reported that one 7.5 cm Panzer-Jaeger had been lost and that three tanks had been knocked out by the 7.5 cm Panzer-Jaeger. This was the last report to mention the 7.5 cm Panzer-Jaeger.

Left and Right: One of the two Pz.Sfl.II completed by Rheinmetall in 1941 on Buessing-NAG HKP 902 chassis with the Maybach HL 45 engine mounted in the rear and sent to the Deutsches Afrika Korps in Libya early in 1942. (NA)

8.8 cm Flak 18 Sfl. auf schwere Zugkraftwagen 12 t (Sd.Kfz.8)

In 1938, the Heereswaffenamt adapted the 8.8 cm Flak 18 to fire at ground targets. For purposes of increasing mobility and ability to quickly get into action, the 8.8 cm Flak 18 took on two forms: a modified trailer mount with an armor shield towed behind an armored 8 to Zugkraftwagen and a self-propelled mount on the chassis of an armored 12 to Zugkraftwagen.

Details of the self-propelled mount were discussed at a meeting held on 16 August 1938.
1. Dr. Porsche explained that Hitler had ordered him to seek methods and ways to accelerate the design and production of the 8.8 cm Flak mounted on the 12 to Zugkraftwagen.
2. Dr. Porsche was briefed by Daimler on the current designs of the 8.8 cm Flak on a trailer drawn behind an 8 to Zugkraftwagen and the 8.8 cm Flak mounted on a 12 to Zugkraftwagen.
3. The following picture resulted from a lengthy discussion on all details of the designs. Dr. Porsche's position was that 12 troop-ready vehicles couldn't be delivered by 30 September 1938. In his opinion, it should still be possible to deliver 10 vehicles by the end of November.

Dr. Porsche proposed that the 12 to Zugkraftwagen be overloaded by 2 tons to allow, among other items, the addition of an armor shield for the gun. Daimler voiced their concerns about overloading the rubber tires on the roadwheels.

Dr. Porsche proposed that devices to block the suspension during firing be dropped. The Versuchsmodell was to be ready by 23 August 1938 for test firing at Kummersdorf to test whether vehicle sway after firing remained within acceptable limits.
4. Dr. Porsche was shown the vehicle. He voiced concerns about the gun's limited depression of only 3 to 4 degrees. It was explained to him that it wasn't possible to redesign the mount in the short time available, and that when depressed to 4 degrees the cradle hit the upper carriage.

Dr. Porsche thought that a traverse arc of 5 degrees to each side was sufficient. However, all others, including the Inspekteur der Fliegertruppen, held the opinion that because of its limited cross-country mobility in overcoming obstacles, situations would easily occur where a gun on the Zugkraftwagen would have to be fired toward the side.

Additional details on the design were recorded in the following report of this same meeting prepared by Dr. Porsche:

The main topic of the meeting was the 8.8 cm Flak mounted on the Daimler DB I 8 chassis. The weight of the chassis alone was 10 tons, which would be increased to about 17 tons by the weight of the gun, ar-

Above: Two 8.8 cm Flak 18 Selbstfahrlafette auf schwere Zugkraftwagen 12t (Sd.Kfz.8) (WJS)

mor, crew, and ammunition. This weight overloaded the chassis by about 2 tons, which necessitated reinforcing the springs and the frame. At first, 6 to 8 mm armor plate was suggested, which from previous experience was not sufficient. However, it is still possible to add sufficiently strong armor protection and remain within the weight limit of 17 tons.

The gun is fired in the direction that the vehicle is facing and can be traversed up to 5 degrees toward either side. This was declared to be acceptable when it was displayed earlier at Jueterbog. However, at this meeting the opinion is that a significantly higher traverse arc is needed to be practical. The firing height of the gun is 2.8 meters above the ground, which can't be reduced without large modification of the chassis. The elevation gear and the mounting of the gun tube allow the gun to be depressed only 3 degrees, which can be increased to 4 degrees after it is modified. Naturally, this arc being very small could result in unfavorable firing conditions if the vehicle itself was on a small slope. Due to the short time available, significant modification of the chassis or the gun is no longer possible.

Because the vehicle displayed at Jueterbog can only be referred to as an Attrappe (mock-up), Daimler-Benz was requested to provisionally complete the vehicle within 6 days so that it could be test fired. These firing trials should prove whether the self-dampening of the counteracting spring mounting is sufficient, or whether shock absorbers must be added. Direct blocking of the springs is not recommended. The engine compartment hood must be reinforced because of the strong muzzle blast, which will be determined during test firing. Daimler-Benz will attempt to improve the field of vision, which was considered to be too restricted in the Attrappenfahrzeug (mock-up vehicle).

After lengthy discussion, Daimler-Benz promised a schedule for completing the first 10 vehicles beginning in mid-November and final delivery by the end of November 1938. Dr. Porsche requested that the remaining 20 vehicles be delivered by the end of this year, which Daimler-Benz will look into. Naturally, this schedule can be met only if the firing trials don't result in any large modifications to the chassis.

Further details were recorded in a memorandum by the Luftwaffe L.C.6 on 28 February 1939:

The plans to knock out concrete machinegun bunkers at short ranges by employing 8.8 cm Flak guns mounted on Selbstfahrlafette or Anhaenger are based on the successful completion of firing trials conducted by the Heeres-Waffenamt. The high muzzle velocity and use of armor-piercing ammunition resulted in destruction of the targets, with 30 percent hits at ranges of 800 to 900 meters.

During the fall of 1938, ten 8.8 cm Flak guns were mounted on 10 to Zugkraftwagen [sic] with sufficient armor protection against S.m.K. ammunition. The ability to engage aircraft was dropped and depression was increased to -4 degrees.

The gun was a modified 8.8 cm Flak 18 with pedestal mount bolted onto the chassis of a 12 ton

Above: Details of the ammunition stowage and gun mounting for the 8.8 cm Flak 18 Sfl. auf 12 to Zugkraftwagen (in this case a Daimler-Benz DBs8 chassis) (WJS)

Zugkraftwagen. Unlike the towed 8.8 cm Flak 18, this self-propelled version had a limited elevation arc of –3 to +15 degrees and therefore the gun could not be used in its original anti-aircraft role. The traverse arc was limited to 151 degrees to the left of center and 151 degrees to the right of center by the upper carriage striking against the armor of the driver's compartment. The telescopic sight for the 8.8 cm Flak 18 was the Flakzielfernrohr 20 E with a meter scale marked on the elevation (range) drum to aid in direct firing.

In the training instructions the crew were advised in a special notes on firing the 8.8 cm Flak 18 auf Sfl.: *The best position for the carriage is when the barrel fires parallel to the direction in which the carriage moves forward or diagonally across the line. If fire has to be opened at right angles to the line of forward movement, care must be taken that the ground beneath it is as level as possible. The recoil from firing in this position causes the carriage to sway. The gun crew must take care to hold on tight.*

A translation of the tactical instructions for the 8.8 cm Flak Sfl. has been included in the Panzer Tracts book entitled "Dreaded Threat" covering the 8.8 cm Flak 18 in its anti-tank role.

Both the DBs8 and DB9 models of the Daimler-Benz 12 ton Zugkraftwagen chassis were used for these self-propelled guns. Both models were quite similar with the DBs8 powered by a Maybach DS O8 rated at 150 horsepower and the DB9 by a Maybach HL 85 TUKRM rated at 185 horsepower. A four-speed transmission with transfer case was geared for a maximum speed of 51 km/hr on roads and 21 km/hr cross-country. Six interleaved large-diameter roadwheels ran on lubricated Zgw 50/400/200 tracks with rubber track pads. Designed to carry 15 tons, most of the weight was supported by the tracks with only 2.4 tons of the load on the front pneumatic tires. As a normal schwere Zugkraftwagen 12 ton (Sd.Kfz.8), the DB9 was 7.10 meters long, 2.40 meters wide, and 2.80 meters high with an adequate ground clearance of 40 cm below the chassis frame. With fuel tanks filled with 250 liters of gasoline, the range on roads was about 250 kilometers. However, fuel consumption off-road was fairly high at 90 to 100 liters per hour.

The 10 modified 8.8 cm Flak 18 Sfl. mounted on the 12 to Zugkraftwagen were issued to the 1.Kompanie/Panzer-Jaeger-Abteilung 8, which took them into action in Poland in September 1939. The success of these self-propelled guns was reported to Daimler-Benz as follows:

Above: An 88 Flak 18 Sfl. auf 12 to Zugkraftwagen with Daimler-Benz DB 9 chassis. (DT)

On the morning of 17 October 1939, Hauptmann Frenzel, commander of the Kompanie of ten 8.8 cm Selbstfahrlafetten, appeared and wanted to have these vehicles examined in the Daimler-Benz assembly plant in Berlin-Marienfelde. At this opportunity, Hauptmann Frenzel raved about the usefulness of these vehicles, especially as artillery and the combination of the gun and Zugkraftwagen.

The impression was left that because of their greater maneuverability and quicker ability to start firing, these Selbstfahrlafetten were vastly superior to the 8.8 cm Flak auf Kreuzlafetten accepted by the Heer. Firing position was changed after 2 to 3 shots, so that not a single loss occurred, because the Polish artillery had no time to engage the target. The troops liked the vehicle so much that it was continuously called on to help, especially to be employed against bunkers. The crews had driven and fired for 21 days, covering a distance of about 6000 kilometers.

The organization of the 1.schwere Kompanie/Panzer-Jaeger-Abteilung 8 was reduced to six 8.8 cm Flak 18 Sfl. in accordance with K.St.N.1146 dated 1 February 1940. Assigned to Guderian's XIX.Armee Korps, the 1.schwere Kp./Pz.Jg.Abt.8 was used to seek out and destroy French tanks in support of the 1. and 2.Panzer-Division during their drive through Belgium and France in May 1940. The 1.schwere Kompanie/Panzer-Jaeger-Abteilung 8 was sent into Russia in June 1941 with its six modified 8.8 cm Flak 18 Sfl. On 29 January 1942 it was renamed Panzer-Jaeger-Kompanie 601 and then on 21 April 1942 again renamed as the 3.Kompanie/Panzer-Jaeger-Abteilung (Sfl.) 559. It was officially organized in accordance with K.St.N.1146 dated 1 November 1941 as the only Panzerjaeger-Kompanie 8.8 cm Flak 18 (6 Gesch.) (mot S) which was to have six 8.8 cm Flak 18 Sfl. on six schwere Zugkraftwagen 12 t (Sd.Kfz.8) als Fahrgestell. On 20 August 1942, Panzer-Jaeger-Abteilung 559 (Sfl.) assigned to A.O.K.2 reported that it had just lost two Sfl.Laf.8.8 m.Zgkw.12t but still had three, of which two were operational. The last three had been lost by March 1943.

Above: Three 88 Flak 18 Sfl. with the 1.schwere Kompanie/Panzer-Jaeger-Abteilung 8, which was sent into Belgium with six on 10 May 1940. (WJS)

10.5 cm K (gp.Sfl.)
previously 10 cm K. Pz.Sfl.IVa

In their fiscal year 1938/39 report Krupp revealed that they had started on the design of a 10.5 cm K. L/52 Selbstfahrlafette (Versuchsgeschuetz). Various proposals for the drive train of a 10.5 cm Sfl. were discussed with Wa Pruef. At first a 320 horsepower engine was considered, and finally a 180 horsepower engine was selected. A wooden model was completed and the Waffenamt awarded a contract for the design and production of two vehicles.

Krupp made further advancements as reported for fiscal year 1939/40. The design of the 10.5 cm Selbstfahrlafette (Pz.Sfl.IVa 1) was completed and contract given to Krupp-Grusonwerk to produce two Versuchsfahrzeuge in armor. The first vehicle is almost complete. The design was worked through for a second model for this vehicle (Pz.Sfl.IVa 2) with six roadwheels. However, the production contract for this second model was rescinded by Wa Pruef 6.

As reported by Krupp for fiscal year 1940/41, two 10.5 cm L/52 Selbstfahrlafette Pz.Sfl.IVa were completed and are being tested by the troops. The acceptance of this type of self-propelled weapon for further production hasn't been determined. Its original purpose as a Schartenbrecher (bunker buster) against the Maginot Line has been overtaken by circumstances. Although they carry heavy armament, they do not have sufficient armor to protect the crews for offensive employment in combat. This invention is in part surpassed by the heavier armor on the VK 45.01 Panzers from Porsche and Henschel for which we developed and are producing the turrets.

Development Details

Having never previously encountered this problem of designing a large-caliber self-propelled gun, numerous ideas and proposals were tried and altered before Wa Pruef 6 and Krupp could arrive at a solution. As with most complicated designs, the end result was a compromise caused by conflicting specifications and limitations on size, weight, speed, and armor protection. Many of the records of project meetings between Krupp engineers and Wa Pruef have survived. These reports provide a comprehensive view of the design as it evolved and a very rare opportunity to learn the actual reasons behind design decisions.

On 25 April 1939, Krupp presented and discussed conceptual drawings W1298 and W1299 with Wa Pruef 6, as follows: *In drawing W1298 the engine was located under the gun so that the barrel overhang was 800 mm shorter than in W1299. This engine location had various disadvantages: poor access to the engine, unfavorable installation of the radiator and cooling air fans, crew encumbered by heat, fumes, and noise from the engine even with shields installed, awkward operation of the transmission, firing height increased to at least 2.00 m, very nose-heavy vehicle, and tendency to rock when the gun was fired.*

In drawing W1299 the engine and cooling system were located behind the gun at the rear of the vehicle and therefore easily accessible. The firing height is reduced to 1700 mm. In previous meetings great value was placed on a lower firing height. Because of a lower seat, the driver can see to the right (underneath the gun). The center of gravity is favorable.

<u>In spite of the disadvantages, Wa Pruef held the opinion that it was better to locate the engine below the gun because the overhang of the gun is shorter.</u>

The torsion bar suspension causes an increase in the height above the belly because the shorter mounts can't be fitted in all positions. In addition the 6-cylinder engine (instead of a V-12) causes a height increase of about 100 mm. Therefore locating the engine below the gun will cause an increase in the firing height and overall vehicle height.

Even with its tactical problems, two-piece ammunition can be used because it has the following advantages:
o *It is possible to load three different charges.*
o *The gun can be served and loaded at all elevations.*
o *The gun can be employed by division artillery.*

<u>By 15 May, Krupp was to prepare and present a new conceptual design drawing with the engine below the gun, a 6-cylinder engine, and torsion bar suspension.</u>

On 2 May 1939, Woelfert and Heerlein from Krupp met with Wa Pruef 6 in Berlin to discuss the new conceptual drawing W1300 with the engine under the gun, as follows:

The radiator could be installed forward to the right in order to obtain better side access to the engine. A different engine was selected because installation of the 6-cylinder Maybach HL 116 engine below the gun results in a firing height of 2100 mm. About 100 mm would be spared by installing the 12-cylinder Maybach HL 120. In addition, there is better access to the cylinders and cylinder heads with the V-12 engine. However, the HL 120 is 200 kg heavier and is to go out of production. Wa Pruef thought that a 6-cylinder Maybach HL 66 engine (rated at 200 horsepower at 3200 rpm) would be sufficient because in comparison to the VK 20.01 the top speed can be reduced from 50 to 40 km/hr. Krupp engineers thought that a power ratio of 10 horsepower per ton was too low for adequate acceleration. Wa Pruef didn't have an opinion. The overall weight could be reduced by installing the transmission

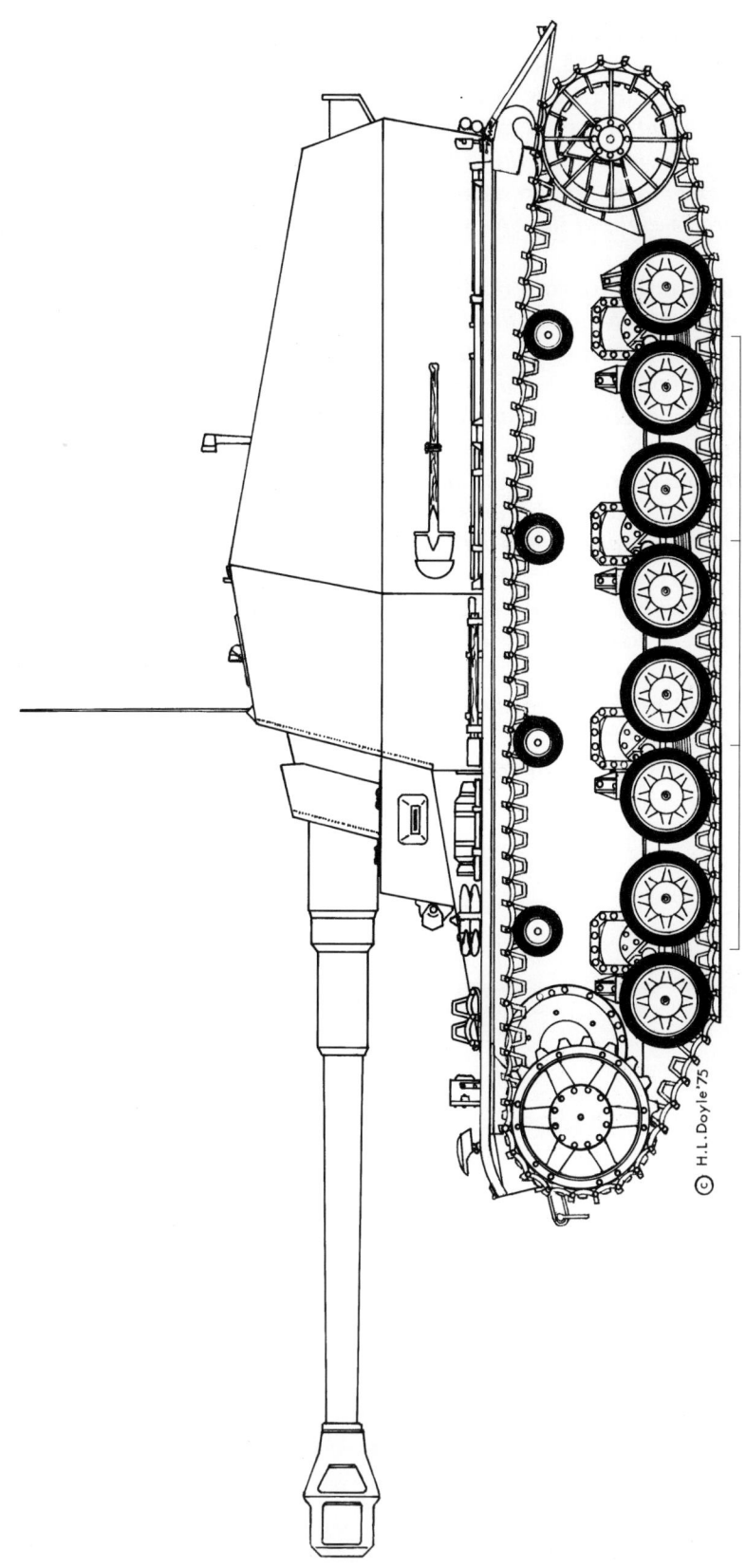

7-33

and steering unit from the VK 9.01 and using dry-pin tracks instead of lubricated tracks with rubber pads.

Recoil of the 10.5 cm Kanone can be reduced from 1000 to 800 mm by adding a muzzle brake. Therefore the gun can be moved 200 mm toward the rear.

In a meeting with Wa Pruef 6 on 15 May 1939, Krupp presented conceptual design drawings W1301 and 1303. In drawing W1301, good access to the engine was achieved by transversely mounting the 300 horsepower Maybach HL 116 and moving the transmission over to the side. The overall weight of the Sfl. was estimated to be 24 metric tons.

Wa Pruef 6 decided in favor of drawing W1303 with a 200 horsepower Maybach HL 66 engine located behind the gun, resulting in a total weight of 22 metric tons with dry-pin tracks. Krupp improved the design by installing the radiator behind the engine with an improved path for the cooling air.

The recoil cylinder and recuperator were to be moved further back to reduce the size of the armor housing. A normal Rundblickfernrohr (panoramic gunsight) was planned for indirect firing. A new Winkelfernrohr (periscopic gunsight) with a longer separation between optics was to be used for direct firing.

Approval was obtained for work to immediately start on the 1:1 scale wooden model of the Sfl.

Krupp informed Wa Pruef 6 that a weight increase couldn't be avoided if a <u>torsion bar suspension</u> *replaced the B.W.-Laufwerk (8 roadwheels paired with leaf springs for the Pz.Kpfw.IV). Also for the torsion bar suspension a recessed section in the side walls is needed so that with an interior hull width of 1.950 m, the overall width of 3.000 m wouldn't be exceeded.*

On 23 May 1939, Krupp presented drawing W1307, which was an enhancement of W1303 with the radiator behind the gun and engine. The hull could be shortened by 40 to 50 cm by moving the radiator from the front to the rear. This would save weight in addition to the weight saved by reducing the length of the track. Since the position of the rear is decided by the gun, a shorter hull would result in the gun being moved 40 cm further forward. Wa Pruef agreed to this.

Using a lubricated track without rubber pads was not possible due to the high specific weight of 85 kg/meter. Wa Pruef will determine if dry-pin tracks are usable with a Schachtell-Laufwerk (interleaved roadwheels). Because 2 metric tons of excess weight can't be saved just by selecting thinner armor, plans must be made to use the normal Z.W.38 torsion bar suspension with 6 roadwheels and dry-pin tracks. This limits the highest speed to 40 km/hr.

On 10 August 1939, Wa Pruef 6 examined and favorably commented on the wooden model of the Pz.Sfl.IV. The following improvements were discussed: To save weight the inside hull width can be re-

Above and Right: One of the two 10 cm K. Pz.Sfl.IVa completed by Krupp-Grusonwerk in January 1941. They were officially renamed 10.5 cm K (gp.Sfl.) in August 1941. (TTM)

duced from 1950 to 1850 mm, brought down to the same width as the B.W. The deck in the crew compartment is to be lowered by 100 mm and the duct for the cooling air exhaust moved to behind the rear wall. This will allow the rear wall to be moved forward and the gun with front wall moved further back. This will also reduce the gun overhang. The roof is to be lowered 80 mm and sloped toward the sides. Later the roof is to be removable in order to remove and reinstall the gun.

On drawing W1316, in places the thickness of the carbon-steel hull can be reduced. The driver's cabin is to be shortened and a pivoting visor with the K.F.F.2 (twin periscopes) installed in the front with 90 mm thick laminated glass blocks behind inset prisms on both sides.

Krupp proposed bins that swing out for stowing cartridge. Some of the projectiles can be stowed upright beside the cooling air duct.

In order not to exceed a weight limit of 20 metric tons, the decision was made to install the HL 66 P engine with a B.W. 8-roadwheel suspension. Krupp proposed the B.W. suspension because it was 430 kg lighter than the Z.W. 6-wheel suspension, took up less interior space, and could be acquired rapidly for the Versuchsausfuerhung. Wa Pruef approved the use of the B.W. suspension and would determine if the VK 9.01 transmission and steering unit as well as the HL 66 P engine could be obtained to meet the schedule. Wa Pruef awarded a contract to design and produce a Versuchsfahrgestell out of carbon-steel (not armor).

On 21 August 1939, the decisions were confirmed that the Maybach HL 66 P engine (rated at 188 hp at 3200 rpm) would be retained along with the transmission and steering unit from the VK 9.02. The 8-roadwheel suspension from the B.W. would be adopted because it had proved successful, was immediately available, and made it possible to reduce the firing height as well as the height for the engine mounts. The top speed was set at 35 km/hr. Details agreed upon included:

Engine - To determine if the engine had sufficient power for a 20 metric ton vehicle, Krupp-Grusonwerk was to perform a test by reducing the output of an HL 120 engine from 265 to 180 horsepower.

Steering Unit - As proposed by Krupp, Wa Pruef will test whether the VK 9.02 steering unit is suitable for the Pz.Sfl.IV because it has a track contact length that is longer than the 9-ton vehicle. In addition, higher turning power is needed because the ratio of the contact length to width is 1:1.46 for the Pz.Sfl.IV in comparison to 1:1 for the VK 9.01.

Suspension - The first two Pz.Sfl.IV were to have the B.W. 8-wheel suspension. Krupp proposed a 6-wheel suspension as shown in Sk 3413, 3414, and 2415 that had the following advantages over the 8-wheel suspension: The Z.W. track with 10 cm tall guide teeth could be used. It also had softer springs, less rolling resistance, larger swing arm deflection both up and

down, no interference in the ground clearance, and better steering ability because the track contact length was 140 mm shorter.

Armor Hull - Just like other Versuchsfahrzeuge, the hull is to be made out of armor immediately so that the armor hull can be tested without losing time.

Krupp gave Wa Pruef a copy of overview drawing W1319. Wa Pruef gave advanced approval for the production of:

2 Pz.Sfl.IV (10.5 cm K18) in armor with 8-roadwheel suspension, outfitted with all military equipment - without the special delivery of a Waffenamt Fahrgestell with 8-roadwheel suspension. Wa Pruef.4 was to order the gun separately.

1 Versuchsfahgestell Pz.Sfl.IV in armor, with 6-roadwheel suspension and a test weight instead of a gun.

On 1 September it was decided that Krupp-Grusonwerk would deliver about 7 metric tons of parts to be installed in the hull, suspension, and final drives and assemble the chassis in automotive running order.

At a meeting in Essen on 4 September 1939, the construction drawing of the Selbstfahrlafette was presented to Wa Pruef 6. It didn't have a turret and the hull and superstructure were constructed as one unit without a connecting flange (as was the usual practice at this time). The frontal armor plates were 50 and 30 mm thick, sides 20, rear 14, and belly 10. The 50 and 30 mm front plates were to be made out of face-hardened armor if possible. Two Versuchsfahrzeuge of this design were to be completed in May and June 1940. Armor hulls would be needed 8 weeks earlier by the end of March and April 1940.

After going through the changes agreed upon, the wooden model of the Pz.Sfl.IV was examined. The main change had been reduction of the width and length. Wa Pruef agreed with the design, and the following was decided:
1. The gunshield front is to be tilted back at about 15 degrees instead of 8.
2. To better protect the openings, a cover is to be installed for the Zielfernrohr and a lid for the extendable Rundblickfernrohres for indirect fire. The center of the roof is to be removable for dismounting the gun. The top middle section of the rear wall is to be cut out 100 mm to allow indirect sighting.
3. Wa Pruef will send an example of the Turm-Spaehfernrohr 320 for the commander.
4. The driver's hatch is to be rectangular and countersunk.
5. Mounts for the Scherenfernrohr (binocular observation periscope) for the loaders are to be located in the forward corner reinforcement.
6. An Entfernungsmessers (range finder) won't be installed.

Confirming earlier information, a third Pz.Sfl.IV with 6-wheel suspension will no longer be considered.

Design Guidelines from Wa Pruef

The following binding design guidelines for the Panzer-Selbstfahrlafette IV (10 cm) for Krupp to follow were dated 19 September 1939:

1. General Description

The Panzerselbstfahrlafette IV (Pz.Sfl.IV, 10 cm) consists of a full-tracked chassis of the Pz.Kpfw.IV type and a schweren 10 cm Kanone (s. 10 cm K.18). The gun is to be mounted on the chassis so that the rear area remains free for the crew to serve the gun. The rear area is to be left open without a roof. Attention is to be paid to weight distribution. The chassis can be lengthened to achieve this. In both thickness and quality the front armor is to be especially penetration resistant. The total weight of the Pz.Sfl.IV may not exceed 20 metric tons.

2. Fahrgestell (chassis)

The 8-wheel suspension is to be adopted from the Pz.Kpfw.IV. The engine is planned to be the 188/200 horsepower Maybach HL 66 P. With the usual installation of the VK 9.02 transmission, the engine is to be displaced so that there is free space in the rear area. The engine must be easily accessible. Cooling air inlet on the side, cooling air outlet to the rear. Under all conditions, the driver is to be seated as far forward as possible in a special driver's compartment. The fuel tank can be located in the front of the chassis.

3. Bestueckung (armament)

The gun to be mounted is the s.10 cm K.18 (10,5 cm L/52). While retaining the gun tube and breech, the gun is to be mounted so that the lowest firing height can be achieved. To achieve this, the recoil cylinder and recuperator with the smallest dimensions are to be located under the gun tube. The recoil cylinder is to be sized to allow a sustained rate of fire of 120 rounds per hour. The shape of the carriage must allow uncomplicated effective armor protection. The gun is to be mounted so that the weight is balanced. The travel lock is to be fastened to the breech block or on the deflector guard. By using a muzzle brake the recoil can be reduced so that firing larger charges will not exceed the allowable braking pressure. The current limit is 11 tons. Mechanical firing device by hand, just like a normal s.10 cm Kanone 18.

Traverse arc about 8 degrees left and 8 degrees right of center. Elevation arc -15 to +10 degrees.

The traverse and elevation gear are to be located on the left side of the gun. Traverse speed of 12 mils per turn of the handwheel = 0.7 degrees. Elevation speed of 12-14 mils per turn of the handwheel.

The gunsight for indirect fire is the artillery Rundblickfernrohr 34 and for direct fire the Selbstfahr-

lafetten-Zielfernrohr (Sfl.Z.F.). For direct fire with Panzergranaten the mount for the Sfl.Z.F. is to be graduated every 100 meters up to a maximum range of 3000 m.

Optical data for the gunner's sights are 2x magnification 20 degree field of view, 10 mm aperture for the Sfl.Z.F.1 and 3x magnification, 20 degree field of view, 8 mm aperture for the Sfl.Z.F.2.

The commander will have a doppelsichtige Turmspaehfernrohr (T.S.F. 3x 20 degrees) for observation. Both of the loaders will have an Artillerie Scherenfernrohr mounted were it won't hinder there loading activities. The driver has a K.F.F.2 and a Grobvisier in front of the view slit. A reserve Sfl.Z.F. and K.F.F.2 is to be stowed in every Pz.Sfl.IV.

5. Sonstige Bewaffnung (secondary armament)

Additional armament for the crew: 4 M.P. with cover and 2 magazine pouches for each. 1 Leuchtpistole in a pouch with 12 Signalpatronen (signal cartridges).

6. Panzerung (armor)

The armor is to be constructed from flat unbent armor plates. The superstructure is to be as open on top as is allowed by the reinforcing plates in the corners. The side walls of the fighting compartment are to be high enough that slightly bent over crew members are protected. Armor thickness of 50 mm front, 20 mm sides, 20 mm roof. As far as possible, face-hardened plates are to be used. This is a requirement for the front plates and the forward superstructure pieces. Penetrations in the frontal armor are to be avoided. The armor housing around the gun slot is to be a simple form

avoiding bent pieces. Above the gunner in the roof is an armor guard that is to cover the entire traverse arc.

<u>Armor shelters are planned to protect the loaders from strafing aircraft.</u> Ammunition bins are to have thin armor covers on top.

The middle part of the roof of the forward superstructure is to be bolted to make it easier to remove the gun. A special hatch is to be installed over the driver. He observes through a 50 mm Drehsehklappe (pivoting vision port).

<u>Tarnaufbauten (camouflage superstructures) give the front a systematic shape. This is achieved with a Fahrerkuppel-Attrappe (fake driver's compartment) to be placed on the right side.</u>

The other crew members have two hatches in the rear wall with steps and handles. It must be arranged that the crew can also leave the vehicle by climbing over the top of the side walls. The superstructure must be covered by a Regenplan (tarp). The deck is to be covered with wood. The armor corners in the fighting compartment are to be cushioned.

7. Munition (ammunition)

In general, two-piece ammunition with three charges will be fired. Both Panzer- and Spreng-Granaten are to be stowed. The amount of ammunition will be decided by the available stowage space. At least 25 rounds are to be stowed. The cartridge stowage must especially prevent jarring. The ammunition is to be covered against dust especially in an open-top vehicle.

8. Besatzung (crew)

The crew consists of five men: 1 Fahrer (driver) to the left vehicle front, 1 Geschuetzfuehrer (commander) right beside the gun, 1 Richtschuetze (left beside the gun), 2 Ladeschuetzen (loaders)in the rear.

9. Befehlsuebermittlung (communication apparatus)

A load speaker and speaking tubes connecting the driver to the commander, driver to the gunner, driver to the left loader, and gunner to the commander.

10. Nachrictengeraet (radio set) is not planned.

11. Sonstiges Zubehoer (miscellaneous equipment)

5 Feldflaschen (canteens), 5 Gasmasken, 5 Kochgeschirre (mess utensils), 5 Gepaecksaecke, 1 Doppelglas (binoculars) by the commander, 1 TSF by the commander, 2 Scherenfernrohre, 1 set of Signalflaggen, 1 Verbandskasten (first aid kit), Reserveglasbloecke, Reserveoptiken (KFF2 and Sfl.Z.F.), and covers for weapons and optical equipment.

Final Changes before Production

On 10 and 11 October 1939, Krupp met with Wa Pruef 6 in Berlin to discuss additional changes in the design of the Pz.Sfl.IVa (10 cm K 18) as follows:

Left & Right:

Two unique features on the Pz.Sfl.IVa were a fake driver's cabin located on the right front for symmetry and an armor shelter located inside at the rear to protect the loader's from strafing aircraft. (TTM)

Wa Pruef 6 informed Krupp that the Maybach-Variorex Schaltgetriebe (semi-automatic transmission) and MAN-Ueberlagerungslenkgetriebe (multiple-differential steering unit) were not to be installed. Wa Pruef 6 suggested the Zahnradfabrik SSG 46 transmission and the MAN Cletrac steering unit (with the smallest turning radius of about 4.5 meters and a clutch blocking part of the torque). This transmission is in mass production and the steering unit had been tested in a Pz.Kpfw.IV.

In order that only mass-produced, proven components be used, an agreement was made to install a normal B.W. steering unit.

Because the engine produces only 188 horsepower and the largest gear ratio in the SSG 46 transmission was only 1:8, the maximum speed was reduced from 35 to 25 km/hr. Slope-climbing ability was to be at least 20 degrees. Krupp immediately contacted Zahnradfabrik to determine if the SSG 46 transmission could be used with a 188 horsepower engine. If Zahnradfabrik had reservations about the SSG 46, Krupp proposed that the A.K.6 S 55 transmission be considered, which had been developed by Zahnradfabrik for a 200 horsepower engine. This transmission was not yet tested, but it was developed under proven fundamental principles.

Further changes are to be avoided or the important drawings won't be completed by the end of November to meet the assembly schedule demanded of Krupp-Grusonwerk.

On 4 November 1939, a final decision was made to install the SSG 46 transmission with a top speed of 27 km/hr. Pz.Kpfw.IV Ausf.E roadwheels were to be used on the Pz.Sfl.IVa1.

Production

On 25 April 1940, Wa J Rue (W.u.G.6) reported on the overall status of armored vehicle production. In addition, to the Panzers already accepted for mass production, Wa Pruef 6 had developed a series of Pz.Kpfw. neuer Bauart that AHA In.6 wanted produced. Among these was the 10 cm Kan (Pz.Sfl.) auf Fgst.Pz.Kpfw.IV of which 2 Versuchsfahrzeug were to be completed in August 1940. AHA intended to order about 100 if these proved to be successful.

Instead of May, June, or August 1940 as originally scheduled, on 5 February 1941 Krupp-Grusonwerk reported that the two Selbstfahrlafetten had been completed in January 1941. At a demonstration for Hitler on 31 March 1941, it was decided that if troop trials were favorable, series production of the 10 cm Pz.Sfl. could begin in the Spring of 1942.

On 13 August 1941, Wa Pruef 4 informed Krupp that the final designation for the Pz.Sfl.IVa was now 10.5 cm K (gp.Sfl.).

Combat Service

On 26 July 1941, Panzer-Jaeger-Abteilung (Sfl.) 521 reported on the experience of its Zug with 10 cm Kanone (Panzer-Sfl.IVa).

Shortly before the start of the campaign against Russia, a Zug with two 10 cm Kanonen (Pz.Sfl.IVa) was assigned as an integral part of the Abteilung. The Zug consists of Kampfstaffel, Munitions-Staffel and Tross. The Kampfstaffel is organized as Zug-Trupp, Geschutz-Staffel with two Geschuetzen, Luftschutz-Kfz.(Sd.Kfz.4) and Fernsprech-Wagen (Nachr.Kfz.15). With this equipment, the Zug is capable of being employed for both direct and indirect fire missions.

The 10 cm K (Pz.Sfl.) was built to penetrate the strongest bunkers in the Maginot Line. It fires two-piece ammunition with the projectile separate from the cartridge. Panzergranaten are fired with a large charge (V_o = 805 m/s), Sprenggranaten with a medium charge (V_o = 635 m/s). 26 rounds are stowed in the Pz.-Sfl. and 80 rounds for each gun on the Munitions-Lkw. The longest range for Panzergranaten is 3,400 meters by direct fire and Sprenggranaten 2,400 meters as direct fire or 10,500 meters as indirect fire.

At 1000 meters the Panzergranate can penetrate every known tank. The traverse arc is 9 degrees to each side, 18 degrees in total.

The chassis for the Panzer-Sfl. is that of the Pz.Kpfw.IV. However, to save room it has a smaller transmission and a 180 horsepower engine. The armor is 50 mm in front, otherwise 20 mm. Thicker armor couldn't be used because of the weight limitation of 21 metric tons.

The crew consist of a Geschuetzfuehrer, a Richt-Unteroffizier, a Fahrer und two Ladeschuetzen.

Combat Action

The Zug was employed in the Vorausabteilung in the fight around and east of Kobryn on 23 June 1941. It fired Sprenggranaten at infantry concentrations and anti-tank guns.

At 2115 hours on 24 June 1941, the Zug supported the attack of two Kradschuetzen-Kompanien crossing over the Szczara. From an open firing position, three artillery positions were engaged with Sprenggranaten at a range of 1100 to 1700 meters and silenced for the duration of the attack. The number of engaged and destroyed artillery pieces couldn't be determined with certainty, because night fell during the action.

On 30 June 1941 in an attack across the Berisina east of Bobruissk, an armored train was engaged by

Right: The interior of the first Pz.Sfl.IVa with gun number RV1 stamped on the breach. It has an Sfl.Z.F.1 gun sight. The Scherenfernrohr 14 Z (scissors periscopes) for the loader was mounted on a pivoting arm. (TTM)

the 10 cm K and driven off. Its sure destruction failed because a bolt broke in the slide for the steering brakes (a frequently occurring problem) so that the Geschuetz came to fire about 5 minutes too late. In this time the armored train pulled back behind a bunch of trees so that it couldn't be taken under fire.

Losses

During the advance toward Sluck, one 10 cm K (Pz.Sfl.IV) exploded. The cause of the explosion couldn't be determined with certainty. It is possible that the explosion was caused by ignition of the cartridges stored in the so-called Luftschutzraum (aerial defense compartment). The crew stated that flames shot out of the Luftschutzraum before the explosion. Probably heat built up in this chamber because of heat from the engine combined with especially high external temperatures on this day.

The crew left the vehicle directly after the first flames shot out, so there was no loss of life. The vehicle still rolled a few paces and then came to a halt. An explosion occurred shortly thereafter. A while later, the shells detonated. The chassis was left behind totally destroyed. The gun appears to be still usable.

Lessons Learned

This Sfl. is not sufficiently maneuverable for employment in a Vorausabteilung (lead unit). The limited traverse of only 18 degrees makes it necessary to turn the entire vehicle to aim at targets. This takes considerable time when done repeatedly especially off-road because of the heavy vehicle and weak engine. In addition, because of its armor layout - only thick in the front and 50 mm lower at the back - it was built for frontal use. The vehicle can be shot into from the side and rear. Employment in a Vorausabteilung requires the ability to quickly engage targets in every direction.

The gun has proven itself capable of supporting an infantry attack from an open firing position by direct fire as in the Szczara crossing. It is not possible to observe our own shots because of large dust clouds raised in front of the gun. The Sfl. must alternatively observe each other's fire or an observation post must be established off to one side, manned by a Richtkreis-Uffz. familiar with the crew. Because of its size, lack of mobility, and large dust cloud raised when firing, in future the Sfl. will only shoot Sprenggranaten by indirect fire.

Up to now, the Sfl. hasn't been employed in its specialized tasks - engaging concrete bunkers by direct fire and engaging heavy enemy tanks in coordination

Above: Both 10 cm Pz.Sfl.IVa were issued to Panzer-Jaeger-Abteilung 521. In comparison to the factory photos, it has a lighter muzzle brake and an external travel lock has been added. (KHM)

with other Panzerabwehrwaffen. Its high penetration ability appears to be suitable for this.

No specific problems have occurred with the engine or transmission. The steering brakes are overstressed. Bolts in the steering slide of one steering brake have torn out three times and the brake bands had to be changed twice because the rivets were overheated.

On 23 August 1941, Panzer-Jaeger-Abteilung (Sfl.) 521 reported on the action of the 10 cm Pak (Sfl.) engaging enemy columns and tanks north of Askolki on 20 August.

Enemy columns marched on a road crossing the front of the firing position. Firing had to be opened at a range of 4000 meters because the next possible firing position was behind about 3-km-wide brushland occupied by the enemy, and other troops weren't available to support an advance through this terrain. Enemy tanks opened fire but remained moving with the column and did not attack the firing position. Therefore, we were forced to engage these at long range. This was an exceptional case, forced by circumstances.

The column was engaged with Sprenggranaten Az and Az mit Verzoegerung (delayed fuze). The range scale was set at (the maximum setting of) 2400 meters and the reticle set higher by feel by the gunner because of the long range - initially thought to be 3000 meters. The lay was improved by spotting hits. The column was broken up by well-placed hits.

The first enemy tank was initially engaged with Sprenggranate Az m V because that was already loaded. The hit was not observed. Later, an impact hole was found beside track marks. Probably this hit damaged the tank's tracks. Three further shots were fired at this same tank with Panzergranate 39 rot. No hit was scored but strikes were observed very close to the tank. The gunner set the range scale for Panzergranaten at 3000 meters (maximum) and then by feel used the reticle to increase the range. Shortly thereafter we observed the crew setting their tank on fire. Scratches from a 10 cm Sprenggranate hit were found at one location on the side armor above the suspension.

Two additional knocked-out tanks that had also been engaged by the Pak 10 cm (Sfl.) were found to the right of the first tank. As had occurred with the first tank, they were first fired at by a Sprenggranate Az m V and then with three Panzergranaten 39 rot. Direct hits were not scored. On the side facing toward the firing position, both tanks had taken a hit that destroyed the suspension.

Because the rail crossing was mined where these tanks were abandoned, it is possible that both ran into mines. However, mine detonations were not observed. Both tanks had hits on the left rear part of the suspension. In fact one 32 ton tank was still firing and didn't stop until after being engaged by the 10 cm Pak (Sfl.). It is obvious that it was set on fire by the crew.

Lessons Learned:

Direct hits and penetration of heavy tanks can't be achieved at 4000 meter range. Shell fragments are likely to damage the suspension disabling the tank and causing the crew to bail out and then destroy their own tank.

Because the Pak 10 cm (Sfl.) can effectively engage columns and enemy concentrations by direct fire with Sprenggranaten at ranges up to 4000 meters, the range scale should be marked up to this range. Armored targets can be effectively engaged at ranges up to 1500 meters.

On 2 September 1941, Panzer-Jaeger-Abteilung (Sfl.) 521 reported on a defensive battle with Russian tanks that occurred on 29 and 30 August.

A Russian heavy tank was destroyed at 100 meter range by the schwere Jaeger (10 cm Sfl.). The commander and gunner had gone ahead on foot and scouted the situation. Having not spotted the Sfl., the Russian tank was knocked out from the flank at close range by a clean penetration. The inside of the tank was completely burnt out.

About 1300 hours, a second Russian heavy tank was spotted near Dewitschi firing at a range of 1000 meters and engaged by the schweren Jaeger. A distinct Panzergranate hit was observed but the tank did not start to burn. The tank remained immobilized until night and didn't open fire again. It must have been towed away by the Russians during the night. On 30 August 1941, a destroyed track link of a Russian heavy tank was found near Dewitschi.

On 30 August, during an attack on Makoff, one heavy Russian tank and two light tanks (T 26) were fired at by the schweren Jaeger at about 1200-meter range. Closer approach was not possible because of enemy anti-tank guns. The Russian heavy tank, obviously immobilized by hits, lay stationary until evening. Both light tanks got away in the forest. The heavy tank was also towed off by the enemy during the night. In addition, with the range scale set at 1500 meters, moving Russian tanks were fired at without success. Also, two Russian heavy tanks were fired at without success with the range scale set at 2000 meters. Altogether a total of 30 Panzergranaten had been expended for both days.

Lessons Learned:

Before every firefight with the schweren Jaeger (10 cm Sfl.), thorough scouting by the commander or gunner on foot or on a motorcycle must occur in order to select the most favorable firing position.

Moving targets can be hit only when the distance traveled to the side is short. The traverse of only 9 degrees to each side is insufficient for hitting targets moving to the side out of the small traverse arc.

At ranges over 1000 meters, the gun shouldn't be fired because a) penetration of the heavy tanks is questionable, b) there is poor observation of where projectiles hit, c) the size of the target appearing in the sight rapidly decreases above 900 meters. This reduction is still greater when the target fired at is lower than the firing position.

An Entfernungsmesser (range finder) is urgently needed for firing Sprenggranaten at ranges over 1000 meters.

The commander's Turm-Spaeh-Fernrohr proved to be unusable. Installation of a Scherenfernrohr in the same location is needed.

Experience on 29 and 30 August shows that in contrast to 20 August, the enemy doesn't always destroy their own immobilized tanks. Instead they are left to be towed away during the night.

The small Panzergranate rot issue for the schwere Jaeger that currently can't be replenished (70 Panzergranate still on hand) doesn't allow tanks immobilized by hits to be fired at until totally destroyed. Therefore, resupply of Panzergranaten 39 rot is urgently needed.

Rebuilt and Back into Action in 1942

Toward the end of 1941, the single surviving 10.5 cm K (gp.Sfl.) was returned to Krupp-Grusonwerk for major repair and rebuild work. When they inquired if the chassis should be converted to 7./B.W. Laufwerk (suspension), they were told on 15 January 1942 to keep the 6./B.W. Laufwerk.

After being rebuilt the surviving 10.5 cm K (gp.Sfl.) was returned to a schwere Panzer-Jaeger-Zug in Panzerjaeger-Abteilung (Sfl.) 521 that was ordered to be combat ready by 1 June 1942. It took part in the Summer offensive on the Eastern Front that eventually headed for Stalingrad. It was not reported as being operational in strength reports from Panzer-Jaeger-Abteilung (Sfl.) 521 from November and early December 1942.

Above: The stag's head tactical symbol for Panzer-Jaeger-Abtielung (Sfl.) 521 is barely visible on this 10 cm K. (gp.Sfl.). It is referred to in their experience reports as a _schwere Jaeger (10 cm Sfl.)_ (BA)

10.5 cm K. Pz.Sfl.IVa

Weapons Data: 10.5 cm Kanone L/52
  Elevation: -15, +10 degrees
  Traverse: 8 degrees L, 8 degrees R
  Gun Sight: Sfl.Z.F.1, 2x, 20 degrees
  Graduated to: 3400 m for Pzgr.
    2400 m for Sprgr.
  Secondary: 3 - 9 mm M.P.

Ammunition: 26 - 10.5 cm Pzgr. & Sprgr.
  576 - 9 mm Patr.f.MP

Crew:
  Commander
  Gunner
  2 Loaders
  Driver

Communication: Speaking Tubes

Measurements:
Length, overall: 7.47 m
Length w/o gun: 5.80 m
Width, overall: 2.86 m
Height, overall: 2.53 m
Firing Height: 1.9 m
Wheel Base: 2.4 m
Track Contact: 3.52 m
Combat Loaded: 22 metric tons

Armor Protection:
  Front: 50/30 mm face-hardened
  Sides: 20 mm
  Rear: 10 mm
  Top: 20/10 mm

Automotive Capabilities:
Maximum Speed: 27 km/hr
Range on Road: 170 km
  Cross Country: 120 km
Grade: 27 degrees
Trench Crossing: 2.3 m
Step: 65 cm
Fording Depth: 105 cm
Ground Clearance: 40 cm
Ground Pressure: 0.82 kg/cm2
Power Ratio: 8.2 HP/ton
Steering Ratio: 1.48
Turning Radius: ?

Automotive Components:
Motor: Maybach HL 66 Pla
  6 cyl. water-cooled
  6.6 liter gasoline
  180 HP @ 3200 rpm
Transmission: Z.F. SSG 46
  6 Forward
  1 Reverse
Steering: Differential
Drive: Front sprocket
Roadwheels: 8x2 per side
Tires: 470/75 Rubber
Suspension: Leaf springs
Track: Dry pin 360 (380)
  Kgs 6111 380/120
Links per Side: 99

Right: The T.S.F. (binocular observation periscope) mounted in the right front roof for the commander was not liked by the unit. Only 26 rounds of 10.5 cm ammunition were carried with cartridge cases stowed separately from the projectiles. (TTM)

4,7 cm Pak(t) (Sfl) auf Pz.Kpfw.I (Sd.Kfz.101) ohne Turm
later renamed Panzerjaeger I

Concerned whether towed Panzerjaeger-Abteilungen with 3.7 cm Pak L/45 would be able to stop attacks by heavy French tanks, like the Char B1 bis, a decision was made to create a makeshift self-propelled gun by mounting the 4.7 cm Pak (t) onto a Pz.Kpfw.I Ausf B chassis. The company of Alkett in Berlin-Borsigwalde was awarded the contract to design and produce a single test vehicle, which was demonstrated to Hitler and representatives from OKH on 10 February 1940.

Production, Organization, and Issue

Converted to self-propelled mounts by removing the wheels, axle, and trails, the 4.7 cm Pak(t) were not confiscated when Germany occupied the remainder of Czechoslovakia in 1939. Skoda was given a new contract by the Waffenamt for 4.7 cm Pak(t) production.

Krupp-Essen was awarded a contract by Wa Pruef 6 to produce 60 Schutzschilde fuer LaS 47 and Deutsche Edelstahlwerke AG in Hannover-Linden was awarded a contract for 72 Schutzschilde fuer LaS (47) to be delivered to the assembly firm Alkett in Berlin-Borsigwalde. Waffenamt inspectors accepted the first 40 4.7 cm Pak (t) auf Pz.Kpfw.I Ausf.B completed in March, 60 in April, and 30 in May 1940. Alkett did not have two guns available to complete the last two in May 1940. There were finally completed and accepted in September 1940 and July 1941.

As recorded as brief notes in the Halder diaries, the unit organization and assignment were still being discussed in February 1940 as follows:
3Feb40 - Pz.I with 4.7 cm gun for Heerestruppen (independent army units)?
5Feb40 - 4.7 cm guns to Heerestruppen or Panzer-Divison? Let's first see the design of mount, demonstration on 10th or 14th Feb.
9Feb40 - 132 Panzerjaeger with 4.7 cm gun (Skoda) on Pz I. 120 to Pz.Div. 12 for Heerestruppen.
10Feb40 - Hitler inspects Pz.I/4.7 cm Sfl. from 1230 to 1400. Decided gun for Heerestruppen not for Pz.Divs
19Feb40 - Panzerjaeger by May (7 Abteilungen). First conversion in March, 1 company per Abteilung.
20Feb40 - Guderian: 4.7 cm Panzerjaeger PzIb can only operate with organic tank repair shops. This makes it necessary to use 4.7 cm for infantry as an ordinary chassis.

On 20 March 1940, the Waffenamt reported that the current plans for issuing the 132 4.7 cm Pak (t) Sfl. were: 1 for Wa Pruef 1, 1 for Wa Pruef 4, 36 to 6 Kp. = 2 Pz.Jg.Abt. by 1Apr40, 54 to 9 Kp. = 3 Pz.Jg.Abt. by 1May40, and 36 to 6 Kp. = 2 Pz.Jg.Abt. by 1Jun40. 6 remain in reserve.

On 31 May 1940, the Waffenamt reported that 18 had been issued to a Heeres Pz.Jaeg.Abt., 81 to 3 Heeres Pz.Jaeg.Abt. (27 each), and 14 to the Ersatzheer. Previous orders called for the issue of 18 to a fifth Heeres Pz.Jaeg.Abt. leaving only 1 out of the total of 132 in reserve.

On 3 March 1940, the Organizations-Abteilung ordered Pz.Jaeg.Abt.521 and 616 to be rearmed with 4.7 cm Pak Sfl. to be combat ready by 15 April 1940. On 26 March 1940 the Pz.Lehr Rgt. at the Panzertruppen Schule in Wunsdorf was ordered to create a Pz.Jaeg.Ersatzkp. (Sfl.) for training personnel for Pz.Jaeg.Abt.(Sfl.) 521 and 616. Two additional Pz.Jg.Abt. (mot S) 643 and 670 were ordered to be converted from towed units by the Pz.Tr. Schule Wunsdorf to be combat ready by 15 May 1940. These were to be organized as:
1 Stab Pz.Jg.Abt. (KStN 1106) with 1 Pz.Kpfw.Ib (Sd.Kfz.101)
1 Nachr.Zug Pz.Jg.Abt. (KStN 1192)
3 Pz.Jg.Kp. (mot S) (KStN 1148 dated 2Apr40) with 4.7 cm Pak (t) (auf Sfl.) and Munitions-Sonder-Anhaenger 32.

On 23 April 1940, Pz.Jaeg.Abt.616 was ordered to be outfitted with 27 4.7 cm Pak Sfl. Pz.Jaeg.Abt.643 and 670 were also issued 27, leaving Pz.Jaeg.Abt.521 as the only unit outfitted with 18 (6 per Kompanie).

Tactical Manual

Tactics to be employed by the schnelle Panzerjaeger were printed in Merkblatt 28/1 dated 6 May 1940:

I. General Tactical Principles

Schnelle Panzerjaeger are unique in their continual readiness to fire, high cross-country mobility, and armor protection. They are therefore qualified to combat enemy tanks by attacking.

Unhindered observation (with an open top) allows early recognition and engagement of enemy tanks which have limited vision.

The Schnellen Panzerjaeger-Abteilungen are Heerestruppen. As a rule they will normally be attached to Panzer-Divisionen, as an exception to other divisions to combat armored vehicles of all types.

The combat unit is the Kompanie. They receive instructions from the Abteilung commander. He must strive to maintain control of this unit and employ them so that all weapons can be fully employed in engaging the enemy. Employment of the entire Abteilung as a single unit can be useful in many situations. Employ-

4.7 cm Pak(t) (Sfl.) auf Pz.Kpfw.I (Sd.Kfz.101) ohne Turm
1.Serie/La.S.47 assembled by Alkett from March to May 1940

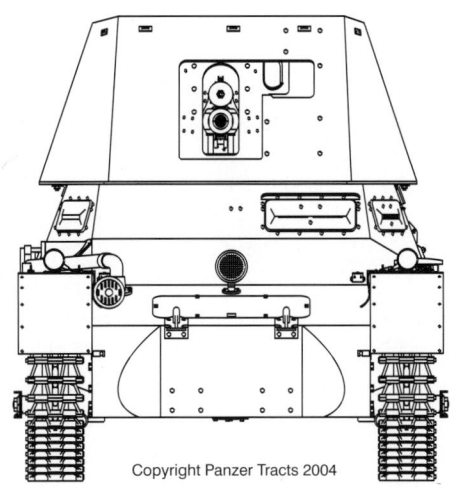
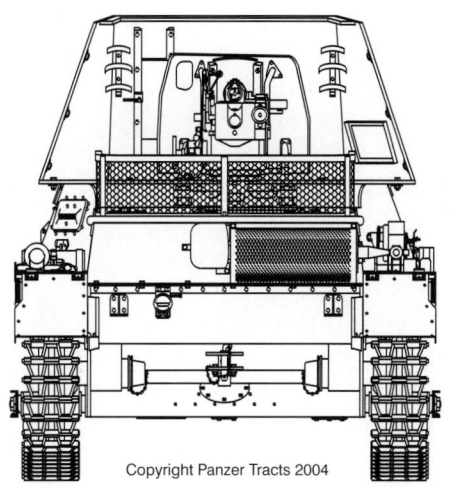
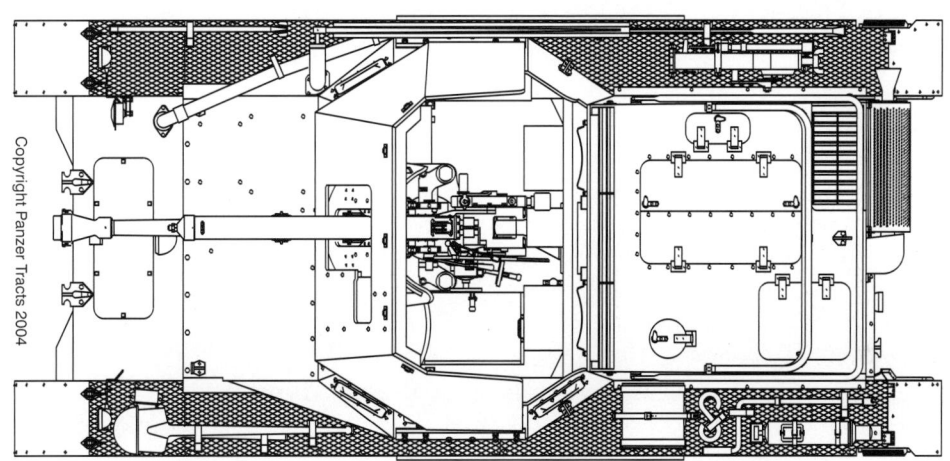
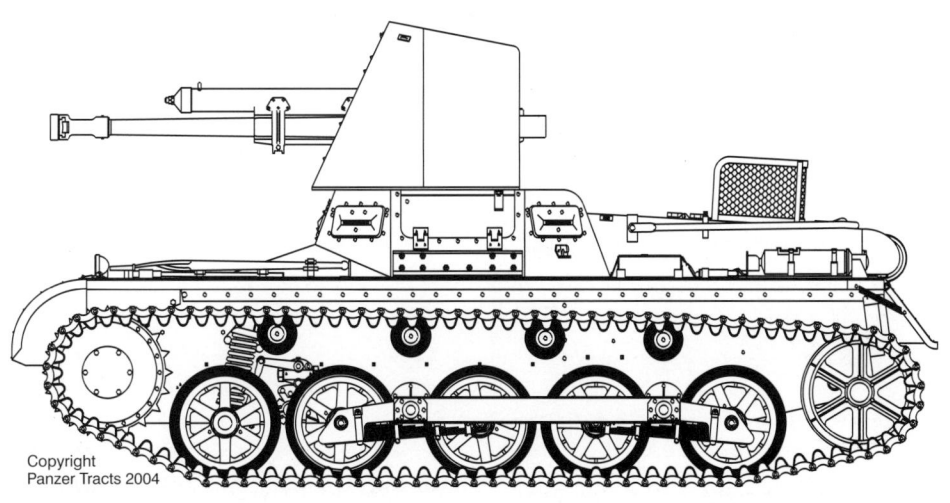

This and Opposite Page:

One of the first series of 132 4,7 cm Pak(t) (Sfl) auf Pz.Kpfw.I (Sd.Kfz.101) ohne Turm completed by Alkett with a five-sided gun shield. Assigned to a Panzer-Jaeger-Ersatz-Abteilung, the number 301 was not a tactical marking from a front line unit. It has been modified by adding a bracket to mount a periscope for the loader, a Notek light and convoy taillight, and also for Tropen (hot climate) employment by cutting slots in the engine hatches that were covered by armor guards.
(TTM)

ment of a single Zug by itself should be seldom.

The decisive elements for successful action are:
a. timely recognition and determining the number and direction of attacking enemy tanks
b. utilizing terrain to advantage
c. rapid handling by commanders at all levels. For this, commanders must be forward.

<u>Combat will be conducted by fire and movement. The purpose of movement is to surprise enemy tanks by bringing them under concentrated fire from unexpected directions and destroy them.</u>

Therefore the Panzerjaeger advance in stages and conduct the firefight while halted. Attacks against the flanks and rear are the most effective.

The task of frontal-advancing Panzerjaeger is to halt the enemy movement by firing. Whether rearward elements are to reinforce the front or be sent against the flanks and rear of enemy tanks is decided by the situation, terrain, and weather.

When engaged by superior enemy fire, the Panzerjaeger are to use their high speed to change positions and continue the firefight.

II. <u>Tactical Employement of the Schnelle Panzerjaeger-Abteilung</u>

The role of the schnelle Panzerjaeger-Abteilung within a Panzer-Division on the march is to protect the front and flanks of the marching columns. Employment of a Zug in the advanced guard can prove useful.

If the Panzerjaeger-Abteilung is attached to an Infanterie-Division, they should march together in the column so that they can quickly be thrown against attacking enemy tanks. The division's own Panzerjaeger-Abteilung (mot Z) can protect the marching columns.

The Panzerjaeger-Abteilung (Sfl.) will often march on the open flank to secure the division's flanks against enemy tanks or motorized units.

When the advancing unit deploys, the entire schnelle Panzerjaeger-Abteilung advances by bounds to knock out attacking enemy tanks as directed by the column commander.

When attacking within a Panzerdivision, as a rule the schnelle Panzerjaeger-Abteilung will be assigned to the Panzerbrigade and attack with it. It has the task of securing the flanks of our Panzers and assisting in the destruction of frontally attacking enemy tanks. The destruction of enemy anti-tank guns is also one of its tasks.

When the objective is reached or the advance is halted to reorganize, the Panzerjaeger are to guard the assembled Panzer units. They will be remain ready for commands from the Brigade commander.

When attacking within an Infanterie-Division or a Schuetzenbrigade, they will concentrate at the Schwerpunkt of the attack in order to destroy attacking enemy tanks. Assignments for cooperation or attachment to the infantry will be determined by the situation.

In exceptional cases, the schnelle Panzerjaeger can be attached to the Infanterie to engage especially bothersome nests of resistance. The decision as to whether the Kompanie commander sends the platoons against individual targets or single platoons work together with the Schuetzenkompanien will be based on the width of the attack strip, the number of targets to be engaged, and closed terrain. One should always strive to quickly concentrate the platoon under cover ready for new assignments from the Kompanie commander as soon as the platoons have completed their combat assignment.

When attacking field or fixed fortifications, the schnelle Panzerjaeger can be employed to fire at bunker embrasures.

After reaching the objective, the schnellen Panzerjaeger are to guard against enemy tank attacks until relieved by the division's Panzerjaeger.

On the defense, the schnellen Panzerjaeger-Abteilungen are to be held in reserve. They won't be sent into action until the direction of the enemy tank attack is clearly identified. Their task is to destroy the enemy tanks that have penetrated into the main battlefield.

After the tank attack is repulsed, the schnellen Panzerjaeger engage the attacking enemy infantry when necessary with Sprenggranaten.

If enemy tanks have broken through, schnelle Panzerjaeger units are to cut off the breakthrough and be sent in to counterattack and destroy the enemy tanks. Exact information about our own mine fields is necessary.

If the situation forces combat to be broken off, together with elements from other units, the schnelle Panzerjaeger-Abteilungen guard against pursuing enemy tank forces.

In especially dangerous situations when our own Panzer forces aren't available, the schnelle Panzerjaeger-Abteilungen with other weapons can be sent in to the attack with a limited objective in order to make it easier for the mass of our own troops to break away from the enemy.

If our own troops retreat, the schnelle Panzerjaeger-Abteilungen takes over concealment and security of the movement. Assignment to the rearguard can be useful.

III. <u>Tactical Handling of the Panzerjaeger I</u>

The crew of the schnellen Panzerjaeger consists of a Panzerjaegerfuehrer (commander) who is also the Richtschuetze (gunner), the Ladeschuetzen (loader), and the Fahrer (driver) who is also the Funker (radio operator)

In conducting a firefight - at the command "Stellung" the Fahrer drives to the next favorable firing position and halts. At the command "Volle Deckung" the Fahrer reverses out of the firing position and moves to the next available cover.

As a rule, driving speed is 15 km/hr. If terrain and the enemy situation require different speed, this is ordered by the Panzerjaegerfuehrer.

The gun may only be fired when halted. Unit concentration should not be lost.

A Panzerjaeger may not fall into enemy hands without the crew and neighboring Panzerjaeger doing everything possible to recover it. If recovery is not possible, destroy the vehicle.

If the Panzerjaeger is in danger of falling into enemy hands, it is to be destroyed. Cleaning rags, burnable material, ammunition, etc. (if possible, soaked with fuel by pulling out the fuel line) are to be ignited in the interior of the vehicle.

IV. Tactics for the Panzerjaeger-Zug

The Zug builds the combat unit. It consists of a Zugfuehrer (platoon leader), three schnellen Panzerjaegern, an M.G.-Trupp, and a Krad (motorcycle), organized and equipped in accordance with K.St.N.1148.

As a rule, the Zug fights with the rest of the Kompanie.

The M.G. is employed for anti-aircraft defense in the assembly area.

The Zug achieves its combat assignments by attacking and firing when halted. Clever exploitation of the terrain by the Fahrer and quickly selecting the firing position by the Zugfuehrer are extremely important to ensure success.

Reverse slopes (in which only the gun tube is above the cover) are the most suitable firing positions.

When the Zug hasn't been spotted by the enemy tanks, it can surprise the enemy by opening concentrated fire. The Zugfuehrer gives the order to fire.

When suddenly encountering enemy tanks at close range, each Panzerjaeger in the Zug self-sufficiently takes up firing positions and opens fire. Contact within the Zug must be maintained.

If the Zug finds itself surprised by superior fire from enemy tanks, the Zugfuehrer quickly orders "Volle Deckung" and retires to a alternative position from which the Zug can engage the enemy under more favorable conditions.

Destruction of the enemy with a few shots and quickly engaging new targets are prerequisites for the success of the Zug.

Troop targets and weapons without armor protection are engaged with Sprenggranaten. If mass targets are encountered during the battle, they are to be destroyed by concentrated fire.

V. Tactics for the Schnelle Panzerjaeger-Kompanie

The Panzerjaegerkompanie consists of a Kp.-Trupp (headquarters), three Zuegen (platoons), the Gefechts- and Gepaecktross (combat and supply train).

Radio silence is rescinded as soon as the unit leaves the assembly area.

As a rule the Kompanie in the first wave advances in the Brietkeil (wide wedge) or Feuerkette (fire chain) formation in order to bring all weapons into effective use. As long as the situation allows, Kompanien in the second or third wave advance in deployed formation.

As long as the terrain and situation allow, the Kompaniefuehrer must strive by attacks to take away the enemy tanks' freedom of maneuver, cause them to go over to the defensive, and destroy them.

When the Kompanie engages in a firefight, the Zuege by themselves engage enemy targets emerging to their front. The Zugfuehrer directs the firefight. The Kompaniefuehrer directs the action by sending short commands by radio to the Zuege or by sending Zuege against the opponent's front or flank.

When the Kompaniefuehrer recognizes that his Kompanie's fire is successful and the opponent starts to pull back, he immediately orders a further attack. The weakening enemy retreat can be hindered by sending Zuege to the flanks and rear.

Switching firing positions during an action is only to occur when effectively engaging the enemy from the present firing position is no longer possible, or when superior enemy fire forces a change.

When assigned flank security, the Kompanie advances in bounds so that at all times they can quickly pull into favorable firing positions from which they can engage enemy tanks attacking the flank.

If a schnelle Panzerjaegerkompanie is assigned to the rear guard, based on the situation and terrain, they can either be concentrated ready to be sent into action by the rear guard commander or individual Zuege can be sent to especially threatened sectors.

If the opponent lays down a smoke screen for an enemy tank attack, the first firing halt for the schnellen Panzerjaegerkompanie should be selected so that the enemy tanks run into their fire as they emerge from the smoke.

If the enemy fires smoke shells onto the Panzerjaegerkompanie firing position, they should pull out to the front or the flank in order to obtain a more favorable position to conduct a firefight.

Data for the schnellen Panzerjaeger I:

Crew: 1 Panzerjaegerfuehrer also Richtschuetze, 1 Ladeschuetze, 1 Fahrer also Funker

Weapon: 1 4,7 cm Panzerabwehrkanone (t) with effective fire out to 1400 meters and the most effective range at 1000 meters.

Ammunition: 74 Panzergranaten and 10 Sprenggranaten on the Panzerjaeger with a further 146 Panzergranaten and 26 Sprenggranaten on the M.G.-Fahrzeugen with Munitions-Anhaenger and with the Gefechtstross.

Armor protection: 15 mm [sic]

Length: 4.30 m *Width:* 2.00 m *Height:* 2.14 m

Marching speed: 30 km/hr on the road, 10-15 km/hr cross-country
Range: 170 km on road, 130 km cross-country
Shortest turning radius: 4 meters (danger of throwing the tracks) Slope: 30° Obstacle: 35 cm
Trench: 1.40 m Ford: 60 cm

Action in the West in May/June 1940

Four Panzerjaeger Abteilung (mot S) took part in the campaign in the West against France. Panzerjaeger-Abteilung (mot S) 521 was assigned to Gruppe von Kleist at the start of the campaign on 10 May 1940. The other three units (Panzerjaeger-Abteilung (mot S) 616, 643, and 670) were sent into action as they became combat ready.

The self-propelled gun proved to be successful, as related in an experience report from the 18.Infanterie-Division: *The 4.7 cm Pak auf Sfl. has proven itself to be very effective against tanks and also against houses when fighting in towns. It had a very large real effect as well as demoralizing effect on the opponent.*

The commander of Panzer-Jaeger-Abteilung 643 (Sfl.) recorded their experiences in getting the unit trained on a new weapon and into action in just over a month. His observations on tactical employment in the following experience report dated 25 July 1940 should be compared to the advice that had been given in Merkblatt 28/1, which was written before there was any actual combat experience with either the weapon or this type of unit.

1. *Training/Outfitting the Unit*

Conversion from a towed 3.7 cm Pak-Abteilung to 4.7 (t) Sfl. occurred in four villages near Wuensdorf from 15 April to 13 May 1940. Scattering the unit over 20 kilometers was very disadvantageous. The long service route from Panzertruppenschule, Panzerlehrregiment and III./Panzer-Lehr-Regiment caused delays. The Abteilung was supported in every way and very well outfitted. Driver training for the drivers and replacement drivers was much too short. The few Fahrschulfahrzeuge (refer to Panzer Tracts No.1-2) allowed only several hours of training for each driver. In the few days available, the Werkstatt, I-Personal, and Panzerwarte (maintenance personnel) could receive only the necessary familiarization but didn't get any experience.

Above: The eye piece for the gunner's sight was angled at 45 degrees for aiming without having to crouch. A lanyard for firing the gun was a standard modification on all 4.7 cm Pak(t) (Sfl.). A Sprechslauch (speaking tube) was added by the unit for the commander to talk to the driver. (BA 54/1526/19)

After two platoon and two live firing exercises - without any Kompanie or Abteilung exercises - the Abteilung had to be combat ready, which it was and loaded for rail transport.

2. Road March Experience

After unloading from the trains, with its inexperienced drivers the Abteilung undertook several long road marches - one lasting 25 hours - in order to join the 3.Infanterie-Division under the III.Armee-Korps at Signy-L'Abbaye. The drivers even managed the night marches without undue problems. Without our own Werkstattzug at least 1 out of 4 Panzer and Panzerjaeger, which had been issued in very poor condition, broke down each day and out of necessity had to be repaired by our inexperienced Panzerwarte. Several Panzerjaeger remained behind for days, but eventually caught up with the unit after the I-Truppe had worked day and night. Even though the Abteilung had taken five trucks full of repair parts from Magdeburg, they still had to continuously get new parts from Luettich, Antwerpen, and Koeln and take the parts to the broken-down Panzerjaeger. The enthusiasm for the weapon and the urge to quickly get at the enemy brought out really distinguished achievements.

3. Further Marches

Coupling the unit with infantry on marches (as was often ordered in contradiction to our advice) ruined the tracked vehicles in a few days. As a result of being overstressed, the clutch steering unit burnt out. Coupling the unit with Panzers for long marches had the same destructive effect. The nose-heavy, overloaded Panzerjaeger Sfl. couldn't sustain the same march tempo as other Panzers. For Panzerjaeger to survive a march in the heat and dust, they can't be driven over 30 km/hr and must halt for a quarter to a half hour after the first 20 kilometers and then after every 30 kilometers in order to cool down, be lubricated, and repaired. Because there weren't any replacement drivers available, 120 kilometers in hilly terrain and 150 kilometers on good smooth roads is the limit for a normal day's march. The distance covered during a night march without lights is less and dependent on the weather and brightness.

The best way is for the Abteilung to move alone. To achieve this it is necessary to attach the unit directly to a Korps with units moving out at a quarter to half hour intervals. Within the unit, tracked vehicles march separately from wheeled vehicles, with the Werkstatt bringing up the rear. Only in this way can almost all vehicles arrive at the objective. The Nachrichtenzug puts up signs with the same Erkennungszeichen (unit recognition symbol) along the march routes to guide broken-down Panzerjaeger that are left behind. In this way Panzerjaeger, left behind for as long as 8 days, always found their way back to the Abteilung. And this at a time when the unit was shifted from division to divi-

Right: One of the first series of 4.7 cm Pak(t) (Sfl.) auf Pz.Kpfw.I issued to one of the four Pz.Jaeg.Abt.(Sfl.) that took part in the campaign in the West in May/June 1940. Dust has all but eradicated the purposefully dark dunkelgrau/dunkelbraun camouflage paint scheme. (GF)

sion - in one period five different divisions in a 4-day period. An astounding achievement.

4. Attachment to Korps and Divisions

The best method is to attach the entire Abteilung to a Korps or Division. Assignment of a single Panzerjaeger-Zug to regiments, battalions, and Abteilungen only led to unanswerable misuse of the Panzerjaeger.

Coupling the entire Abteilung or a single Panzerjaeger- Kompanie with an M.G.-Batallion, Aufklaerungs-Abteilung or Radfahrschwadron as a Vorausabteilungen (lead unit) resulted in great success when the Panzerjaeger being employed as normal Panzers or Sturmgeschuetz had sufficient flank and rear protection.

Employing the Panzerjaeger-Abteilung Sfl. alone (without 3 to 4 Radfahrern or infantry troops assigned to every Panzerjaeger) to guard an open flank can lead to the loss of the entire Abteilung were it to be subjected to a strong infantry attack.

The 4.7 (t) Abteilung was frequently ordered about, here and there, in response to the psychological fear of tank attacks and the supposed ineffectiveness of the 3.7 cm Pak in dealing successfully with enemy tanks. This even sometimes occurred when a few old armored cars appeared and infantrymen then called in fantastic reports about tank attacks. This once led to the Abteilung being pulled away from a sector where an actual attack from several 34-ton tanks was being waged.

5. Effectiveness of the 4.7 cm Pak(t)

The 4.7 cm Panzergranaten are very good against 45 to 50 mm thick armor at ranges up to 500 meters - sufficient up to 600 meters. Up to 1000 meters the tank's tracks can be hit and Pzgr. ricocheted through the belly. It is very good against M.G. nests at 1000 meters - at longer ranges this target is too small for the telescopic gunsight. The trajectory of the Pzgr. is flat - out to 2000 meters! The demoralizing effect on the enemy caused by on-rolling Panzerjaeger was really outstanding - above all when they were firing Pzgr. and high-explosive rounds. Our Panzerjaeger crews have unlimited faith in their gun.

6. Observation

Observation is very bad. You have to look over the shield to observe to the front, resulting in Kopfschuesse! In effect the crew are blind when attacking in villages or against street barricades, M.G. nests, and individual tanks. The Jaegerfuehrer must always keep the target in the gunsight - which is very difficult to do when moving. Looking inward from the side, the Ladeschuetze must continuously pay attention to his gun and loading activities. The Fahrer has to concentrate on his route and can't shout loud enough to be heard by the Jaegerfuehrer during combat. A courageous enemy soldier can easily knock out the crew by throwing a hand grenade out of his foxhole close to the side or rear of the Panzerjaeger. Warnings sent by shortwave radio message from the Kompaniechef, are seldom heard by the Jaegerfuehrer due to combat noise and the excitement in battle.

7. Armor Protection

The Abteilung knows that this Panzerjaeger Sfl. was an emergency birth and is the first of a wonderful weapon of the future. The chassis has insufficient armor. The French 2.5 cm anti-tank gun penetrates even at the farthest ranges! The Panzershild (superstructure) is also penetrated by everything above 7.92 mm S.m.K. When hit by anti-tank shot or artillery shells, pieces of armor are torn out which endanger the crew and rip open terrible wounds.

The openings by the gun tube and by the gunsight are too large. In spite of this, there isn't a very small sight hole for direct aiming in close combat without using the telescopic gunsight. A thicker, at least somewhat better protective shield to the sides and rear with vision devices must be developed.

The trained Panzerjaeger recognizes and copes with these deficiencies and would never exchange this outstanding weapon for the towed 3.7 cm Pak.

8. Ammunition

The ammunition is outstanding. The stowage boxes swell so that after a few days the rounds don't fit correctly. Then the boxes won't shut. Three different ammunition types on a Panzerjaeger can easily lead to the wrong round being grabbed in the heat of combat. In future 50% Pzgr. and 50% Sprgr. should be carried.

9. Room in the Panzerjaeger

Space is much too restricted. Rations and other necessary items roll about with ammunition on the floor when crossing rough terrain resulting in a jumbled mess after a few days. The Panzerjaeger crews don't see their Waeschebeutel I (wash kit) and blankets for days. A box or sack fastened to the track guards is needed to store the most necessary supplies.

10. Befehlspanzer (Pz.Kpfw.I Ausf.B)

The Pz.Kpfw.I is much too small and has insufficient armor. The commander must have room to move about when he performs his tasks of leading, calling on the radio, firing the machineguns, and observing. He also needs the security of sufficient armor. While moving, the driver or commander can hardly get to the radio set to switch from transmitting to receiving or continuously adjust the radio to keep on frequency. A larger and better armored Befehlswagen must be provided without regard to the interests of our own Panzer units.

Right: The stag's head tactical emblem for Pz.Jaeg.Abt.(Sfl) 521 on a first series 4.7 cm Pak (t) (Sfl.) auf Pz.Kpfw.I (NA)

11. Height of the Panzerjaeger

If the French artillery had always been observant, wanted to fight, and weren't demoralized and shaken by the Stukas, Panzer and Stuttgarter Sender, our high superstructure would have been dangerous. The firing height of the 4.7 cm Pak (t) Sfl. is good when standing in grain fields, behind hedges, straw stacks, etc. Because the depression arc of the gun is insufficient to compensate for the chassis being tilted back on reverse slopes, it can't take up "hull down" positions behind or on the flanks of hills. Only front slope positions are possible and ensure success. Moving to a concentrated position behind ridges or in depressions 1 hour after daybreak will almost always be tactically correct - the Panzerjaeger can always counterattack. If the Panzerjaeger were remained in exposed positions during the day, they would be quickly spotted and knocked out by enemy artillery before they could fire against enemy tanks.

Therefore, specifications for a new Pak Sfl. design should demand a high firing height but a low all-round superstructure with medium thickness armor.

When the Jaegerfuehrer is firing, communication with the Fahrer is rarely successful even when employing the known technique of kicking him in the back. Because of the limited traverse of the gun, the Panzerjaeger must be quickly turned in the direction of new targets appearing on the flanks or side. This coordination must be practiced in training if it is expected to occur during combat and under enemy fire.

12. Tank Attack Warning Signals

The blue and violet smoke cartridges that were intended for warning about tank attacks are usually not available. Yellow is also not available, and on the defensive red smoke signals indicate Sperrfeuer (artillery barrage). The Abteilung advises all divisions to which they are assigned to change the tank attack warning to 2 red signals fired close together.

Without in any way attempting to diminish their well-earned title as Koenigen der Waffen (king of the service branches), the Abteilung experience is that when individual infantrymen are asked about agreed upon tank attack signals, the usual response is: "It was once... but today we don't know." Then when a unit is attached to five different divisions in 4 days, sometimes without being able to speak to the Ia (operations officer) or anyone else in the division, he concludes: "Rely only on yourself. Send out all Kraeder (motorcycles) and Funkstellen (radio sets mounted on cars) that the Abteilung possesses as far as possible in all directions toward the enemy. Don't hold back even a single messenger or command set."

13. Supply Train Weapons

When fighting in and behind a Panzer-Division, the opponent routinely comes out of hiding after the Panzers have passed by. The Gefechts- and Haupttrosse are now insufficiently armed with a few rifles and pistols to cope with this situation and must be well

armed with M.G. and M.P.38. When defending a single broken-down Panzerjaeger, the M.P.38 has been great at successfully repulsing attacks from all sides.

14. Zugkraftwagen 1 ton for M.G.34

The Zugkraftwagen 1 ton (Sd.Kfz.10) is an outstanding vehicle. An old gas pipe erected behind the middle partition was used to mount the M.G.34 for anti-aircraft defense while on the move in all marches and proved to be very useful. The M.G.34 didn't fit the Stellage (mount) delivered from the ordnance depot and had to be modified by the Abteilung.

<u>Concluding Remarks</u>

The Panzerjaeger are proud and happy to claim possession of this outstanding weapon of the future - in place of the 3.7 cm Pak towed by old Westwall trucks. They wish nothing better than to be sent to England, Afrika, or to the coast to attack enemy warships. "Vorwaerts im Panzerjaegergeist!"

<u>Further Production and Additional Units</u>

On 19 September 1940, Krupp-Essen received a contract to deliver 70 Schutzschilden fuer LaS 47 to Alkett in Berlin-Borsigwalde. This contract was modified on 15 October 1940. Only 10 were to be sent to Alkett, with the remaining 60 sent to Kloeckner-Humboldt-Deutz A.G. Werk Poll, Koeln-Deutz for assembly. The Waffenamt reported production and acceptance of 10 4,7 cm Pak (t) auf Pz.Kpfw.I Ausf.B in November, 30 in December 1940, and 30 in February 1941.

Additional units were raised/converted to Panzerjaeger-Abteilung (mot S). Pz.Jaeg.Abt.169 (renamed 529) was the fifth Abteilung to be issued 4.7 cm Pak (t) Sfl. At first it had two Kompanien with 9 each; later a third Kompanie was added. On 28 October 1940, Pz.Jaeg.Abt.605 was the sixth Abteilung ordered to rearm with 4,7 cm Pak (Sfl). A Panzer-Jaeger-Kompanie 4.7 cm Pak (Sfl.) was also created for the Leibstandarte SS-Adolf Hitler, which had nine 4.7 cm Pak(t) (Sfl.) on 15 March 1941. A Panz.Jaeg.Kp., 4,7 cm Pak (Sfl.) (9 Gesch.) KStN 1148 v.1.2.41 was also ordered to be created for the Lehr Brigade (mot.) 900 on 15 April 1941.

<u>Operation Barbarossa starting 22 June 1941</u>

The confirmed assignment of Panzerjaeger-Abteilungen (mot S) with 4.7 cm Pak (t) Sfl. auf Pz.Kpfw.I (Sd.Kfz.101) at the start of Operation "Barbarossa" on 22 June 1941 was:
521 to XXIV.A.K., Pz.Gr.2, H.Gr.Mitte
529 to VII.A.K., 4.AOK, H.Gr.Mitte
616 to Pz.Gr.4, H.Gr.Nord
643 to XXXIX.A.K., Pz.Gr.3, H.Gr.Mitte
670 to Pz.Gr.1, H.Gr.Sued

Above: A second series 4.7 cm Pak(t) (Sfl.) auf Pz.Kpfw.I with a seven-sided gun shield followed by one from the first series with a five-sided gun shield. (WR)

4.7 cm Pak(t) (Sfl.) auf Pz.Kpfw.I (Sd.Kfz.101) ohne Turm
2.Serie/La.S.47 completed by Alkett in November 1940 and Kloeckner-Humboldt-Deutz A.G. in December 1940 and February 1941

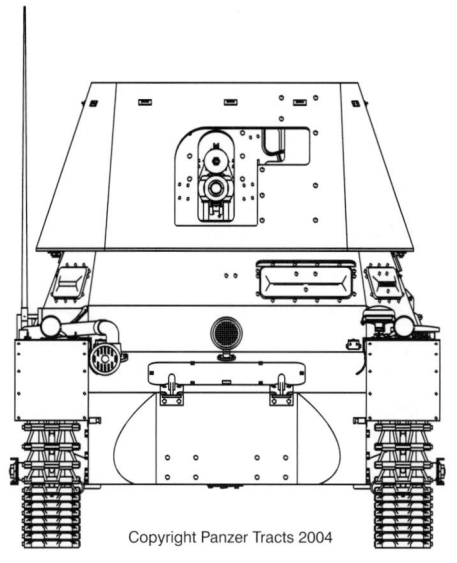
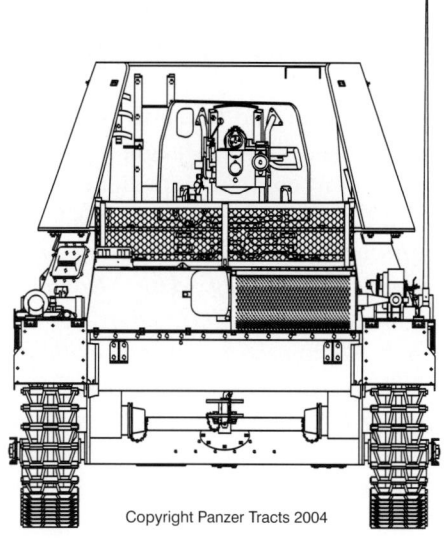
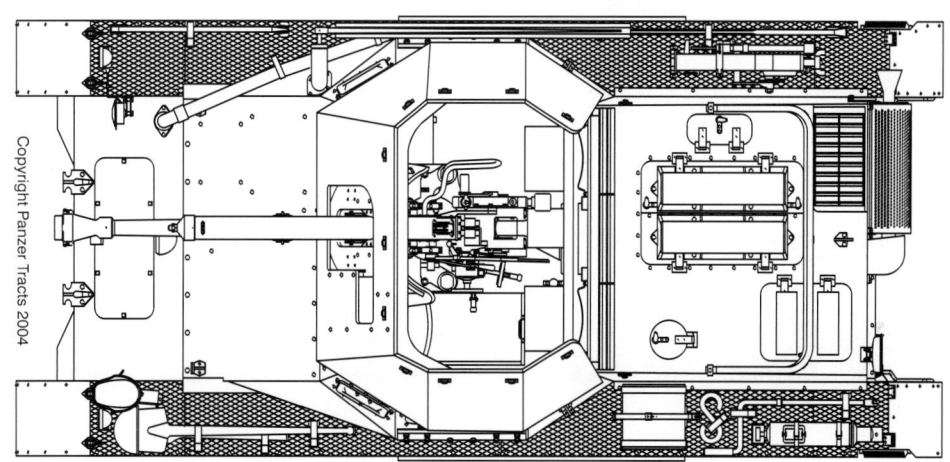
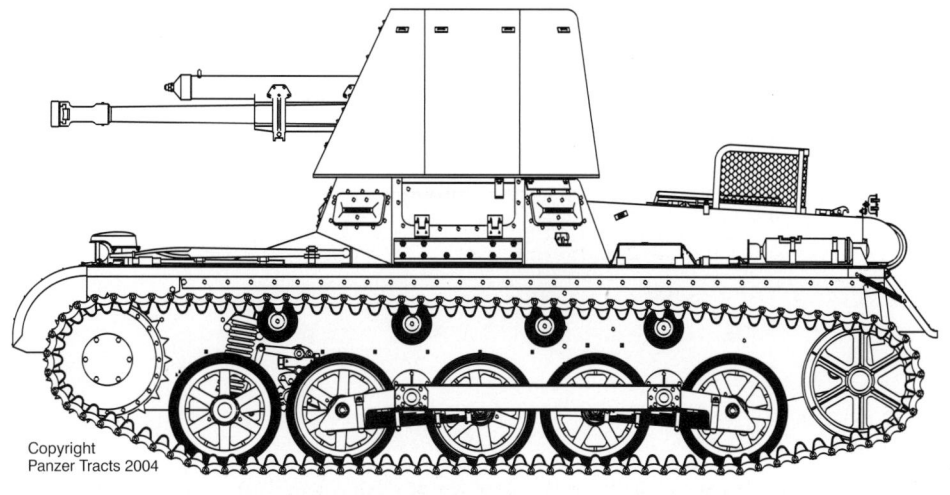

Pz.Jaeg.Abt.(mot S)605 had been sent down to Libya with the 5.leichte Division.

Pz.Jaeg.Abt.529 reported starting the campaign with 27 Pz.IB mit 4,7 cm (t) and 4 Pz.IB. They had lost 4 Pz.IB mit 4.7 cm (t) by 27 July 1941 and still had 23 operational. On 23 November 1941 they reported that 14 of the remaining 16 Pz.IB mit 4.7 cm (t) were still operational.

The first unit that had taken the 4.7 cm Pak (t) into action in France, Panzer-Jaeger-Abt.(Sfl.) 521 reported on their new combat experience in Russia during the period from 3 to 18 July 1941 as follows:

For fighting power the Abteilung has three Kompanien and a schweren Zug. The 2.Kompanie has been assigned to the 1.Kavallerie-Division since 16 June 1941. Each Kompanie is outfitted with nine 4.7 cm Pak(t) auf Fahrgestell des Pz.-Kampfwagens Ib, one Pz.-Kampfwagen Ib for the Komp.-Fuehrer, three M.G. auf ungepanzertem 1 to-Zgkw., and additionally two M.G. were issued without the associated troops to man them.

The 4.7 cm Pak(t) auf Fahrgestell des Pz.-Kampfwagens Ib was produced within one week in March 1940 as a makeshift solution to defend against heavy French tanks. Because of its high penetration capability and rapid rate of fire, during the campaign in France they proved to be very effective in combat against the heavily armored French tanks.

Up to now during the campaign against Russia, the Abteilung hasn't been engaged in its main role of fighting enemy tanks because enemy tank attacks haven't occurred in the 3.Panzer-Divisions sector. However, we did manage to knock out one armored car.

<u>Employed like Sturmgeschuetz</u>

The Abteilung was used exclusively to support infantry attacks and for security tasks. A Panzerjaeger auf Selbstfahrlafette can't be used as a substitute for a Sturmgeschuetz, as can be seen from the following comparison. The Sturmgeschuetz has 50 mm (frontal) armor, all-round armor protection with a closed roof, a low profile and firing height, and a 7.5 cm gun. The 4,7 cm Pak(t) auf Sfl. has 14 mm armor, a Schutzschild open to the top and rear, too high a profile, and a 4.7 cm gun.

The effective range for the 7.5 cm Pak(t) is 1000 to 1200 meters with a maximum range of 1500 meters. When attacking an enemy position equipped with anti-tank guns and artillery, as occurred near Mogileff and Rogatscheff, because of its high superstructure that presents a good target for artillery and anti-tank guns, the Panzerjaeger is destroyed before it can get into action.

When large shells explode closeby, fragments penetrate the thin armor, as occurred near Rogatscheff. Russian 4.5 cm anti-tank guns already penetrate at 1200 meters range. The 1.Kompanie lost 5 out of the 10 Kampffahrzeugen in such actions, of which only two could be repaired.

In contrast, employment was very successful against the enemy equipped only with machineguns, as in crossing the river at Beresina and at Tschaussy; direct fire against machinegun nests is especially effective and helped the infantry advance. However, employment

Above: A second series 4.7 cm Pak(t) (Sfl.) auf Pz.Kpfw.I loaded onto a Sonderanhaenger 115 being towed by a 12 to Zgkw. (Sd.Kfz.8) on the Eastern Front. (BA 266/72/35)

in the first infantry wave can't be considered because the crews must stand up and are not protected against rifle and machinegun fire from the side.

<u>Defensive Security</u>

In contrast to towed Pak, because of their high superstructure and limited traverse arc, the Panzerjaeger must be held back under cover and drive forward into scouted firing positions in response to an enemy attack. This occurred with a Panzerjaeger-Zug attached to Abwehrgruppe Frank near Petrowitschi, where in cooperation with two Krad-Schuetzen-Zuegen (motorcycle infantry platoons), an attack of about 100 Russians supported by artillery and mortar fire was repulsed without loss. In contrast to this, the Abteilung was ordered to drive into firing positions near Rogatscheff, attracted enemy artillery fire, and lost one dead and six wounded without themselves being able to respond.

Without infantry being attached, the Abteilung cannot successfully take on security tasks. They cannot repulse infantry units attacks with the 4.7 cm Pak. There are only 41 combat troops in a Panzerjaeger-Kompanie and 11 in a Zug. An infantry-style defense can't be mounted by our own forces.

<u>Scouting</u>

Scouting difficult terrain can only be accomplished by our own troops. The infantry are not in a situation to determine the passability of trails for the vehicles in the Panzerjaeger-Abteilung.

<u>Anti-tank Mines</u>

The Abteilung must be briefed in advance when they are to operate in a sector where mines have been reported and want to have a Pionier-Gruppe attached. Near Tschaussy, a Panzerjaeger-Kompanie ran into mines within the defensive sector and lost two Kampffahrzeuge to mines. As determined later, the mine fields were already reported but the Kompanie wasn't informed.

<u>Mechanical Problems with Pz.Kpfw.Ib Fahrgestell</u>

Roadwheels without the rubber vulcanized to steel rings lose the rubber tires after about 200 kilometers, while they last about 500 kilometers with steel rings. Sand causes especially high wear on track pins and track links. The starters don't work in 90 degree Celsius heat. This is very disadvantageous while under fire in heavy terrain.

In another experience report dated 2 September 1941, Panzer-Jaeger-Abteilung (Sfl.) 521 reported on their defense against Russian tanks as follows:

On 29 August 1941 during a Russian attack on Schostenkij, a 4.7 cm Panzerjaeger (Sfl.) was destroyed at a range of 120 meters by a heavy Russian tank with Christie suspension. Two officers and a sergeant from another unit were standing on the Panzerjaeger and gave the Geschuetzfuehrer contradictory instructions. This interfered with his ability to calmly assess the situation and take action. They didn't get a shot off, even though the turret of the enemy tank was not initially aimed at them.

On 30 August 1941, in Woronesh-Ost a 4.7 cm Panzerjaeger (Sfl.) surprisingly encountered a Russian BT tank (10-11 tons) as it rounded a bend. The Russian tank drove at high speed directly at the Panzerjaeger. The Fahrer immediately halted and the Geschuetzfuehrer fired two shots. Two crewmen abandoned the tank after the first shot. After the second shot, the burning tank rammed into the Panzerjaeger and tore a piece out of the hull. The Panzerjaeger succeeded in backing up before the flames spread to it.

These actions with 4.7 cm Panzerjaeger (Sfl.) both show that leaders from other units cannot ride into combat with the Panzerjaeger or decisions by the Geschuetzfuehrer will be hindered. This is the only explanation as to why the Panzerjaeger at Schostenkij didn't open fire against a stationary Russian tank.

On 5 May 1942, Panzerjaeger-Abteilung (Sfl.) 521 reported that they still had five Pz.IB mit 4.7 cm (t) in one Kompanie and three surviving Pz.IB but expected to have the other two Kompanien outfitted shortly with 18 Pz.38t mit 7.62 cm Sfl. and 12

Left: This first series 4.7 cm Pak(t) (Sfl.) auf Pz.Kpfw.I is still utilizing the original antenna mount for the Funksprechgeraet "a" radio set. (BA L 28248)

Pz.Kpfw.I ohne Aufbau fuer Munitions-Transport. In 1942, Panzerjager-Abteilung (Sfl.) 670 also had one Kompanie with 4.7 cm Pak (t) Sfl. along with two companies converted to 7.62 cm Pak Sfl. Panzerjager-Abteilung (Sfl.) 616 still had all three companies outfitted with 4.7 cm Pak(t) Sfl. through at least the Fall of 1942. Panzerjaeger-Abteilung (Sfl.) 529, still with two 4.7 cm Pak(t) Sfl., was disbanded with its 1.Kompanie renamed 3.Kp./Pz.Jg.Abt.27 for the 17.Panzer-Division.

Operation Sonnenblume Starting February 1941

Panzerjaeger-Abteilung 605 with 27 4.7 cm Pak(t) Sfl. was sent to Libya with the 5.leichte-Division (mot), arriving by ship in Tripoli on 18 to 21 March 1941. Several Panzer-Jaeger were lost in combat by June 1941 and three replacement Panzer-Jaeger arrived in Tripoli on 2 October 1941 (two others had sunk on the ship "Castellon").

At the start of Operation "Crusader" on 18 November 1941, Panzerjaeger-Abteilung 605 had 27 4.7 cm Pak(t) Sfl., lost 13 during the campaign, and still had 14 at the end of the year. They received four replacements in January 1942 and had 17 4.7 cm Pak(t) Sfl. at the start of Operation "Venezia" on 25 May 1942. Three additional replacements were sent in September/October 1942. The unit still had eleven 4.7 cm Sfl. auf Pz.Ib at the start of the major British offensive at El Alamein on 23 October 1942. Two more replacements arrived in early November 1942, too late to get into action at El Alamein.

An assessment of the good and poor characteristics of the 4,7 cm Pak(t) Sfl when employed in Nord Afrika was reported in July 1942 as follows:

The good accuracy of this weapon was especially commented on. Usually a hit is obtained with the first shot at ranges up to 1000 meters. Penetration ability is too low for the necessary combat ranges in the desert. It was reported to be the best of the smaller caliber self-propelled guns. The chassis (Pz.Ib with Maybach engine) is too weak. The engine is overtaxed.

The guard on the breech opening handle doesn't provide sufficient protection. The firing mechanism is impractical. It is desired that a trigger be mounted on the traverse handle with pressure applied in the firing direction.

Increasing the gunsight reticle to a range of 4000 meters for Sprenggranaten is desired. The reticle in the sight should be illuminated at night.

The Funkgeraet "a" doesn't have a long enough range and is too susceptible to damage in the present mounting. The battery is not sufficient.

Springs in the suspension are continuously breaking.

All-round armor is desired - but won't occur due to the weight. At least reinforce the frontal armor.

In one case, three Mk.II (Matilda infantry tanks) were penetrated at a range of 400 meters by 4.7 cm Pz.Gr.40 (tungsten core ammunition). It usually penetrates 60 mm of armor. Therefore, a small percentage of these rounds is desired. The 4.7 cm Pzgr.36(t) will not penetrate a Mk.II at 600 to 800 meters. But the crews will abandon the tank because fragments spall off the armor on the inside.

Left: This second series 4.7 cm Pak (t) (Sfl.) auf Pz.Kpfw.I, assigned to the first platoon leader in the 1.Kompanie of Pz.Jg.Abt.605, has an antenna mount bolted to the left rear support for the gun shield. (BA L 18 458)

4.7 cm Pak(t) (Sfl.) auf Pz.Kpfw.I (Sd.Kfz.101) ohne Turm

Weapons Data: 4.7 cm Pak(t)
  Elevation: -8, +10 degrees
  Traverse: 17.5 degrees L, 17.5 R
  Gun Sight: Z.F. 2x, 30 degrees
  Graduated to: 1400 m?

Secondary: 1 - 9 mm M.P.

Ammunition:
74 - 4.7 cm Pzgr.Patr.
10 - 4.7 cm Gr.Patr.
192 - 9 mm Patr.f.MP

Crew:
Commander/Gunner
Loader
Driver/Radio Operator

Communication: Fu.Spr.Ger."a"

Measurements:
Length, overall: 4.42 m
Length w/o gun: 4.42 m
Width, overall: 2.06 m
Height, overall: 2.14 m
Firing Height: 1.72 m
Wheel Base: 1.67 m
Track Contact: 2.4 m
Combat Loaded: 6.4 metric tons
Fuel Capacity: 146 liters

Armor Protection:
Gun shield: 14.5 mm proof against S.m.K.
Chassis: Refer to Panzer Tracts 1-1

Automotive Capabilities:
Maximum Speed: 40 km/hr
Avg. Road Speed: 25 km/hr
  Cross Country: 10-15 km/hr
Range on Road: 170 km
  Cross Country: 115 km
Grade: 30 degrees
Trench Crossing: 1.40 m
Step: 37 cm
Fording Depth: 60 cm
Ground Clearance: 29.5 cm
Ground Pressure: 0.48 kg/cm2
Power Ratio: 15.6 HP/ton
Steering Ratio: 1.44
Turning Radius: 4 m

Automotive Components:
Motor: Maybach NL 38 Tr
6 cyl. water-cooled
3.8 liter gasoline
100 HP @ 3000 rpm
Transmission: ZF F.G.31
  Reverse: 3.7 km/hr
  1.Gear: 5.0 km/hr
  2.Gear: 10.4 km/hr
  3.Gear: 19.3 km/hr
  4.Gear: 30.8 km/hr
  5.Gear: 40.0 km/hr
Steering: Clutch-Brake
Drive: Front sprocket
Roadwheels: 5x1 per side
Tires: 530/72 Rubber
Suspension: Leaf springs
Track: Dry pin 260 (275)
Kgs 67 280/90

Right: A second series 4.7 cm Pak(t) (Sfl.) auf Pz.Kpfw.I with a seven sided gun shield flanked by two from the first series with five sided gun shields. They all still have the 1.4-m-long antenna rod for the Funksprechgeraet "a" in the pivotal mount on the right side. (WS)

4.7 cm Pak (t) (Sfl.) auf Fgst.Pz.Kpfw.35 R 731(f)

On 23 December 1940, AHA/AgK (In 6) authorized Wa Pruef 6 to initiate a design project for installing the 4.7 cm Pak (t) onto the Pz.Kpfw.R35 to create a Panzerjaeger (Sfl.). Originally, these were intended to be assigned to Infanterie-Divisionen that were not motorized. Wa Pruef 6 awarded the contract for detailed design and completion of an Entwicklungsstueck (development piece) to Alkett in Berlin-Borsigwalde. This first example of a 4.7 cm Pak (t) (Pz.Sfl.) auf Fgst.Pz.Kpfw.R35(f) with a carbon-steel (not armor) superstructure was completed on 8 February. Demonstrated to Hitler on 31 March 1941, it was announced that a series of 200 had been ordered, at least 30 would definitely be completed in April, and the entire series completed by August.

The contract for assembly of 200 4.7 cm Pak(t) auf Pz.kpfw.R35(f) ohne Turm was awarded to Alkett, which began work in February 1941 to meet a schedule for delivery of 30 per month starting in March 1941. Delays occurred, with the first 93 reported as completed in May 1941, followed by 33 in June, 5 in July, 22 in August, 28 in September, and the last 19 in October 1941. Out of the total of 200, 174 were completed as Geschuetz-Fahrzeuge and 26 as Fuehrungs-Fahrzeuge (command vehicles). Fuehrungs-Fahreuge had an M.G.34 in a ball-mount instead of a 4.7 cm Pak (t).

As ordered on 27 February 1941 by the OKH Organizations-Abteilung, Panzer-Jaeger-Abteilung 559 and 561 were to be rearmed with 4.7 cm Pak(t) (Sfl.) beginning in April. Each Kompanie was to consist of three Zuege with three Sfl. in each Zug. Panzer-Jaeger-Abteilung (Sfl.) 559, 561, and 611 were outfitted with the first 81 4.7 cm Pak (t) (Sfl.) auf Pz.Kpfw.R35(f) and 12 Fuehrungs-Fahrzeuge that had been completed in May/June 1941. On 19 June 1941, the training unit 3.Kp./Pz.Jg.Ers.Abt.43 was ordered to be converted to a Pz.Jg.Ers.Kp. (Sfl.) 35R (f) responsible for replacement personnel for Pz.Jg. Abt. (Sfl. 35 R) 559, 561, and 611. It was issued four 4.7 cm Pak (t) (Sfl.) and six Pz.Kpfw.-Fahrgestelle 35 R (f) that were to be sent by rail from the Panzer-Lehr-Regiment in Wunsdorf to the 3.(Sfl. 35 R) Kp./Pz.Jg.Ers.Abt.43.

At the start of Operation "Barbarossa" on 22 June 1941, Pz.Jg.Abt.(Sfl.35R) 559 was assigned to the LVI.A.K. in Heeresgruppe Nord, Pz.Jg.Abt.(Sfl.35R) 561 to A.O.K.9 in Heeresgruppe Mitte, and Pz.Jg.Abt. (Sfl.35R) 611 to the XXXXVII.A.K. in Heeresgruppe Mitte.

On 3 July 1941, both Pz.Jg.Abt.559 and 611 were ordered to be rearmed from Sfl. to 3.7 cm Pak.

As recorded in the XXXXVI.A.K. war diary on 5 July 1941, the commander of Pz.Jg.Abt.611 re-

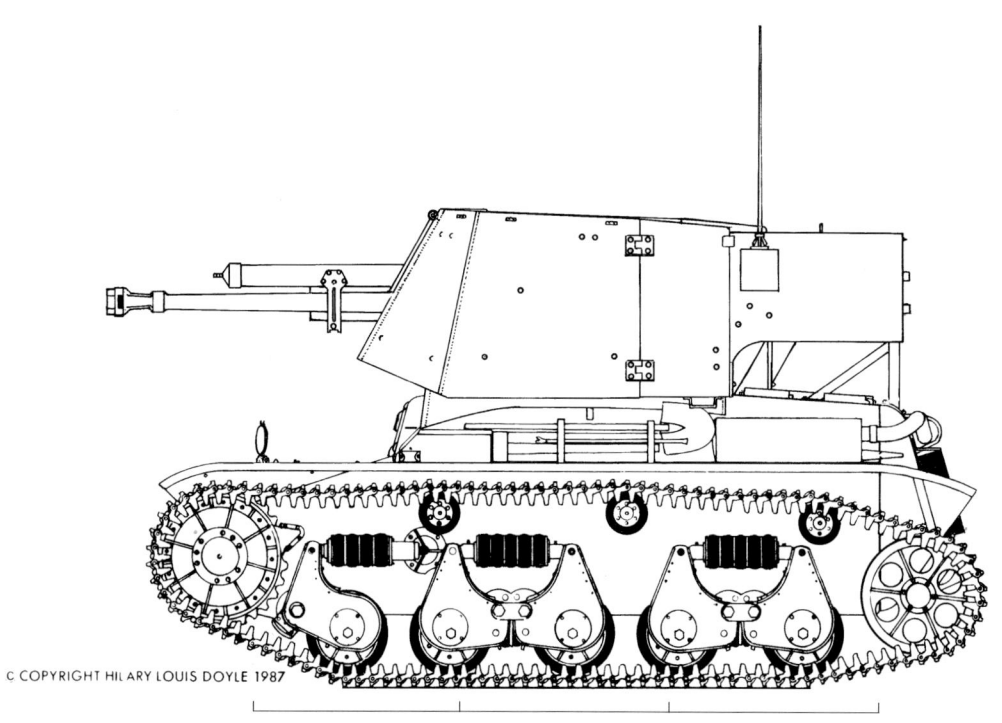

4.7 cm Pak(t) (Sfl.) auf Fgst.Pz.Kpfw.35 R 731(f)

ported that the French vehicles had all completely broken down due to mechanical failure in the first days of the campaign. Two Kompanien reoutfitted partially with German Pak and partially with captured Russian anti-tank guns are fully combat ready; the third Kompanie was planned to be ready in about 1 or 2 days. On 4 July 1941, it was recorded in the A.O.K.9 war diary that Pz.Jg.Abt.(Sfl.) 561 was currently being rearmed and would remain in army reserve in Grodno until it had achieved mobility.

Panzer-Kompanie 318, issued 10 Pz.Kpfw. 35 R 731 (f) als Sfl. mit 4.7 Pak (t) and 2 Fuehrungs-Fahrzeuge, was sent to the rueckw. H.Geb.Sued (rear army zone in southern Russia). In an experience report from February 1942, the unit reported a catastrophe had occurred due to weak engines, poor roads, and weather conditions. These vehicles were reported to be unusable for any employment on long stretches, especially in the winter. At -10 degrees C the engines wouldn't start, the grease froze so that roadwheels wouldn't turn, the parallel ribs on the tracks caused the vehicles to slide sideways off icy roads, and the Funksprechgeraet "a" failed at -20 degrees C due to low power from the batteries.

In spite of these failed attempts at front-line service, 4.7 cm Pak(t) auf R35 were kept in service, mostly in the West. Strength reports from late June to December 1943 record that in the West there were 6 with Pz.Jg.Abt.657, 3 with Pz.Rgt.100, 8 to 10 with 18.Lw.Feld.Div., 8 with 156.Res.Div., 10 with 171.Res.Div., 10 with 191.Res.Div., 2 with 243.Inf.Div., 2 with 266.Inf.Div., 11 with 343.Inf.Div., 2 with 346.Inf.Div., 6 with 348.Inf.Div., 2 with 371.Inf.Div., 5 with 709.Inf.Div., 2 with 711.Inf.Div., 2 with 712.Inf.Div., 2 with 716.Inf.Div., 10 with Festungsstamm-Abt.1/XXV, and 24 on the channel islands of Guernsey and Jersey with the 319.Inf.Div. A very high percentage of the Pz.Kpfw.35R chassis were operational with these garrison units - 85 out of 96 in late June 1943, and 88 out of 92 in December 1943.

In common usage, various German army reports written from 1941 to 1943 kept switching the name between R35 and 35R. However, the "official" designation from D50/12 is Panzerkampfwagen 35 R 731 (f).

As reported on 30 July 1941, Wa Pruef 6 had given Alkett a contract to design and produce a single 5 cm Pak 38 auf R 35 (f) that was intended for Panzerjaeger units in non-motorized Infanterie-Divisionen. With an armor superstructure 25 mm thick on the front and 20 mm on the sides and manned by a crew of three, it was predicted to weigh 11.5 metric tons and have a maximum speed of 20 km/hr. It is not known whether the single 5 cm Pak 38 auf R 35 (f) that had been scheduled for delivery in August 1941 was actually completed.

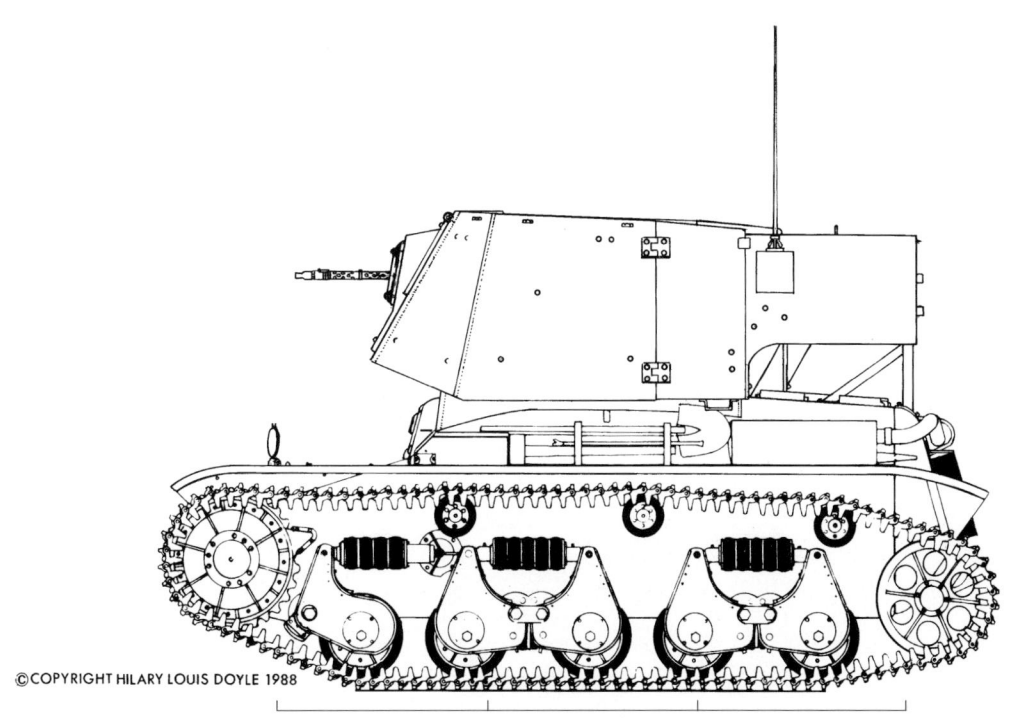

Fuehrungs-Fahrzeuge with Fgst.Pz.Kpfw.35 R 731(f)

Above and Left:
A company of 4.7cm Pak(t) (Sfl.) auf Fgst.Pz.Kpfw.35R on a training exercise. A 1 ton Zgkw. (Sd.Kfz.10) assigned to each platoon for transporting extra ammunition and towing an ammunition trailer had a M.G.34 mounted for anti-aircraft defense.
(WR)

Above and Right:
A platoon of 4.7 cm Pak(t) Sfl. auf Fgst.Pz.Kpfw.35R out on a patrol in the dunes.
(HLD)

7-65

Above and Left:
The Fuehrungs-Fahrzeug (command vehicle for company and battalion commanders) on the Fgst.Pz.Kpfw.35R had a Kugelblende 30 (machinegun ball mount for 30-mm-thick armor) bolted onto an armor housing in place of the 4.7 cm Pak(t) (Sfl.). (WR)

4.7 cm Pak(t) (Sfl.) auf Fgst.Pz.Kpfw.35 R 731(f)

Weapons Data: 4.7 cm Pak(t)
  Elevation: -8, +10 degrees
  Traverse: 17.5 degrees L, 17.5 R
  Gun Sight: Z.F. 2x, 30 degrees
  Graduated to: 1400 m?
Secondary: 1 - 9 mm M.P.

Ammunition: ? - 4.7 cm Patr.
192 - 9 mm Patr.f.MP

Crew: Commander/Gunner
Loader
Driver

Communication: Fu.Spr.Ger."a"

Measurements:
Length, overall: ? m
Length w/o gun: 4.20 m
Width, overall: 1.85 m
Height, overall: ? m
Firing Height: ? m
Wheel Base: 1.56 m
Track Contact: 1.56 m
Combat Loaded: 11 metric tons
Fuel Capacity: 168 liters

Armor Protection:
Gun shield: 25 mm front, 20 mm sides
Chassis: 32 mm front, 40 mm sides/rear

Automotive Capabilities:
Maximum Speed: 20 km/hr
Range on Road: 130 km
  Cross Country: 80 km
Grade: 25 degrees
Trench Crossing: 1.40 m
Step: ? cm
Fording Depth: 60 cm
Ground Clearance: 32 cm
Ground Pressure: 1.36 kg/cm2
Power Ratio: 7.7 HP/ton
Steering Ratio: 1.00
Turning Radius: 8.5 m

Automotive Components:
Motor: Renault
4 cyl. water-cooled
5.8 liter gasoline
85 HP @ 2200 rpm
Transmission: 4 forward, 1 reverse
Steering: Cletrac
Drive: Front sprocket
Roadwheels: 5x1 per side
Rubber tires
Suspension: Rubber discs
Track: Dry pin 260/70
Links per Side: 123

Right:
A 4.7 cm Pak(t) (Sfl.) auf Pz.Kpfw.35 R with the 1.4-m-long antenna rod for the Funksprechgeraet "a" mounted on a flexible base. The radio set Fu.Spr.Ger."a" had a voice transmission range of 2 to 3 km when stationary, reduced to 1 km when on the move. (HLD)

Pz.Sfl.Ia
5 cm Pak 38 auf gp.Mun.Schlepper

Directly following the victory in the West, on 5 July 1940, In 6 authorized Wa Pruef to develop a leichte Panzerjaeger for the Luftlandetruppen (airborne troops) and Infanterie Divisionen that weren't motorized. Wa Pruef 6 awarded a contract to Rheinmetall-Borsig for detailed design of a 5 cm Pak 38 L/60 mounted in an armor superstructure on a Munitions-Schlepper chassis designed by C.F.W. Borgward in Bremen. To save weight, the specification called for the gunshield/superstructure to have Schottenpanzer (spaced armor) with two 4-mm-thick plates in front and two 3-mm-thick plates on the sides. Both the frontal and side armor was to be sufficient to defeat hits from 7.92 mm S.m.K. (steel core AP bullets).

The chassis was a converted VK 3.02 Munitions-Schlepper. Powered by a 2.3 liter Borgward engine, rated at 50 horsepower, the maximum speed for the 4.5 metric ton Panzerjaeger was 30 km/hr. A crew of three manned this Panzerjaeger with a commander (also serving as a gunner) along with a loader and a driver.

Panzerprogramm 41, dated 30 May 1941, planned for 3144 leichte Panzer-Jaeger (Pz.Sfl.) auf VK 3.02 Fgst. to be produced for Panzer-Jaeger-Abteilungen in normal Infanterie-Divisions and Luftlande Truppen. However, production wasn't planned to start until 1943, with 100 completed by 1 April 1944 and 200 completed by 1 April 1945.

Two Versuchsgeraete (trial pieces) were scheduled to be completed after July 1941. These two Versuchsgeraete in armor had been completed by 1 July 1942 and sent to the front for troop trials.

On 8 August 1942, 19.Panzer-Division reported the Fgst.Nr. of the two 5 cm Sfl. Borgward to OKH. An experience report from Panzer-Jaeger-Abteilung 19 on the modified Pz.Sfl. (5 cm Pak 38) auf Mun. Schlepper was presented to OKH on 18 September 1942. Unfortunately a copy of this report was not retained with the attachments to the division's war diary.

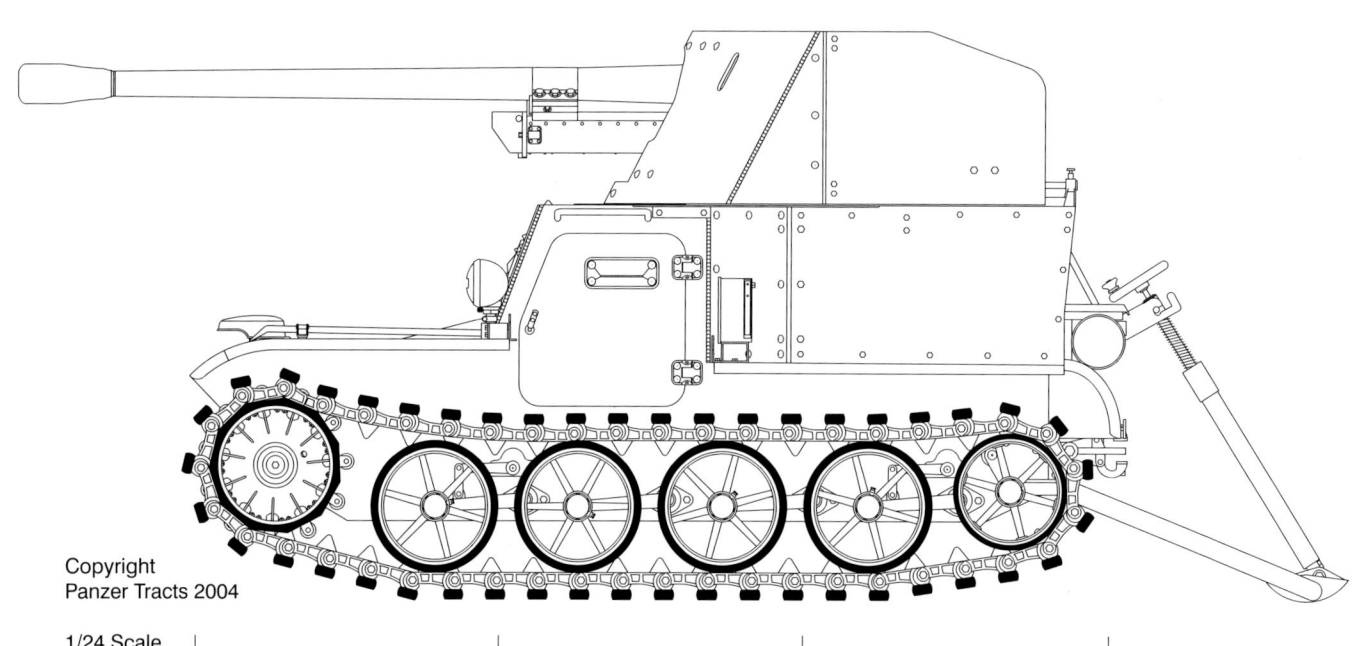

Copyright Panzer Tracts 2004

1/24 Scale

Above and Left:
One of the two trial Pz.Sfl.Ia produced by Rheinmetall by mounting a modified 5 cm Pak 38 L/60 gun on a Borgward gepanzerter Munitions-Schlepper (VK 3.02). (RC)

7-70

This and Opposite Page: One of the two trial Pz.Sfl.1a completed by Rheinmetall using a gepanzerter Munitions-Schlepper (Fgst.Nr. 330033) that had been assembled originally by Borgward in 1941. (RC)

Pz.Sfl.Ic
5 cm Pak 38 auf Pz.Kpfw.II Sonderfahrgestell 901

On 5 July 1940, In 6 had authorized Wa Pruef to develop a leichte Panzerjaeger for Panzer-Divisionen and Infanterie-Divisionen (mot.). Wa Pruef 6 awarded contracts to Rheinmetall-Borsig to design the 5 cm Pak mounted in an armored superstructure on a Pz.Kpfw.II n.A. VK 9.01 chassis from M.A.N.

In 1941, the chassis from the VK 9.01 series was powered by a Maybach HL 45 engine, rated at 150 horsepower with a drive train geared to achieve a top speed of 70 km/hr. The more powerful Maybach HL 66 engine, rated at 188 horsepower, was installed in the VK 9.03 chassis that was predicted in 1942 to have a top speed of 50 to 60 km/hr. With 30 mm thick frontal and 20 mm side armor, the Panzerjaeger weighed 10.5 metric tons. The 5 cm Kanone L/60 was a modified Pak 38 with the breech and carriage adopted from a 5 cm Kw.K. The Pz,Sfl.Ic was manned by a crew of four.

Panzerprogramm 41, dated 30 May 1941, called for the production of 1200 l.Pz.Jaeger (Pz.Sfl.5 cm) auf VK903 Fgst. plus another 828 for Schuetzen (motorized infantry) and Aufklaerungs-Abteilungen (reconnaissance battalions). However, no prediction was made as to when production would occur. In July 1941, it was reported that two Versuchsstuecke (trial pieces) in armor were scheduled to be completed after September 1941.

On 10 March 1942, OKH ordered the assignment of the two 5 cm Pak Sfl. to the 3.Zug/Panzer-Jaeger Kompanie 601 to replace the 8.8 cm Sfl. lost in 1941. The 1.Kp./Pz.Jg.Abt.8 had been renamed Pz.Jg.Kp.601 on 29 January 1942 and then renamed 3.Kp./Pz.Jg.Abt.(Sfl.)559 on 21 April 1942. In the strength report dated 20 August 1942, these two Pz.Jg.Sfl. 5 cm Fgst.Ic (one operational) were reported as present with Panzer-Jaeger-Abteilung (Sfl.) 559 under A.O.K.2 on the Eastern Front.